CONTENTS

July 18, 2002

Page

Opening Statements

Witnesses

Appendix 1: Answers to Post-Hearing Questions

Appendix 2: Additional Material for the Record

NASA WORKFORCE AND MANAGEMENT CHALLENGES

HEARING

BEFORE THE

SUBCOMMITTEE ON SPACE AND AERONAUTICS

COMMITTEE ON SCIENCE

HOUSE OF REPRESENTATIVES

ONE HUNDRED SEVENTH CONGRESS

SECOND SESSION

JULY 18, 2002

Serial No. 107–85

Printed for the use of the Committee on Science

Available via the World Wide Web: http://www.house.gov/science

U.S. GOVERNMENT PRINTING OFFICE

80–686PS WASHINGTON : 2003

For sale by the Superintendent of Documents, U.S. Government Printing Office
Internet: bookstore.gpo.gov Phone: toll free (866) 512–1800; DC area (202) 512–1800
Fax: (202) 512–2250 Mail: Stop SSOP, Washington, DC 20402–0001

COMMITTEE ON SCIENCE

HON. SHERWOOD L. BOEHLERT, New York, *Chairman*

LAMAR S. SMITH, Texas
CONSTANCE A. MORELLA, Maryland
CHRISTOPHER SHAYS, Connecticut
CURT WELDON, Pennsylvania
DANA ROHRABACHER, California
JOE BARTON, Texas
KEN CALVERT, California
NICK SMITH, Michigan
ROSCOE G. BARTLETT, Maryland
VERNON J. EHLERS, Michigan
DAVE WELDON, Florida
GIL GUTKNECHT, Minnesota
CHRIS CANNON, Utah
GEORGE R. NETHERCUTT, JR., Washington
FRANK D. LUCAS, Oklahoma
GARY G. MILLER, California
JUDY BIGGERT, Illinois
WAYNE T. GILCHREST, Maryland
W. TODD AKIN, Missouri
TIMOTHY V. JOHNSON, Illinois
MIKE PENCE, Indiana
FELIX J. GRUCCI, JR., New York
MELISSA A. HART, Pennsylvania
J. RANDY FORBES, Virginia

RALPH M. HALL, Texas
BART GORDON, Tennessee
JERRY F. COSTELLO, Illinois
JAMES A. BARCIA, Michigan
EDDIE BERNICE JOHNSON, Texas
LYNN C. WOOLSEY, California
LYNN N. RIVERS, Michigan
ZOE LOFGREN, California
SHEILA JACKSON LEE, Texas
BOB ETHERIDGE, North Carolina
NICK LAMPSON, Texas
JOHN B. LARSON, Connecticut
MARK UDALL, Colorado
DAVID WU, Oregon
ANTHONY D. WEINER, New York
BRIAN BAIRD, Washington
JOSEPH M. HOEFFEL, Pennsylvania
JOE BACA, California
JIM MATHESON, Utah
STEVE ISRAEL, New York
DENNIS MOORE, Kansas
MICHAEL M. HONDA, California

SUBCOMMITTEE ON SPACE AND AERONAUTICS

DANA ROHRABACHER, California, *Chairman*

LAMAR S. SMITH, Texas
JOE BARTON, Texas
KEN CALVERT, California
ROSCOE G. BARTLETT, Maryland
DAVE WELDON, Florida
CHRIS CANNON, Utah
GEORGE R. NETHERCUTT, JR., Washington
FRANK D. LUCAS, Oklahoma
GARY G. MILLER, California
MIKE PENCE, Indiana
J. RANDY FORBES, Virginia
SHERWOOD L. BOEHLERT, New York

BART GORDON, Tennessee
NICK LAMPSON, Texas
JOHN B. LARSON, Connecticut
DENNIS MOORE, Kansas
ZOE LOFGREN, California
SHEILA JACKSON LEE, Texas
BOB ETHERIDGE, North Carolina
MARK UDALL, Colorado
DAVID WU, Oregon
ANTHONY D. WEINER, New York
RALPH M. HALL, Texas

BILL ADKINS *Subcommittee Staff Director*
ED FEDDEMAN *Professional Staff Member*
RUBEN VAN MITCHELL *Professional Staff Member*
CHRIS SHANK *Professional Staff Member*
RICHARD OBERMANN *Democratic Professional Staff Member*
AMANDA PARSONS *Staff Assistant*

NASA WORKFORCE AND MANAGEMENT CHALLENGES

THURSDAY, JULY 18, 2002

HOUSE OF REPRESENTATIVES,
SUBCOMMITTEE ON SPACE AND AERONAUTICS,
COMMITTEE ON SCIENCE,
Washington, DC.

The Subcommittee met, pursuant to call, at 10 a.m., in Room 2318 of the Rayburn House Office Building, Hon. Dana Rohrabacher [Chairman of the Subcommittee] presiding.

Chairman ROHRABACHER. Good morning. This hearing will examine the Administrator's vision for NASA, and NASA's legislative proposals for what is necessary to realize this vision. But before we have opening statements and before we have testimony, I would like to offer the full Science Committee Chairman Boehlert the opportunity to express his position and his thoughts about NASA and the challenges we face. Sherwood Boehlert, who has been very, very fair and honest and hard working, and has offered a great deal of positive leadership to this committee, and we appreciate him being with us.

CHAIRMAN BOEHLERT'S OPENING STATEMENT

Chairman BOEHLERT. Thank you very much, Mr. Chairman. I appreciate those kind words. Unfortunately, I can't stay at this hearing because I have to participate in the Intelligence Committee joint inquiry on 9/11, but I did want to let everyone know some of my thoughts on the issues before us today. Also, I am also pleased to see Administrator O'Keefe, and I wanted to ensure that he, in particular, understood where we are on these issues.

Like Administrator O'Keefe, I believe that there are few things more important for NASA than ensuring that it has the best possible workforce in the future. With the expected retirement bulge and the competitive market for top scientists and engineers, NASA may be left without the workforce it needs in the coming years. That is a situation we must avoid. NASA has to have the cream of the crop as it did at its inception.

So one thing I wanted to see this committee do this year is to move forward with some proposals that would ensure that NASA has the people it needs. NASA has brought forward its own package, which we are reviewing, and I will carefully study the testimony that we hear today. I want to understand better, for example, concerns raised by the unions. But I do believe we can come up with proposals that will help NASA and gain widespread support.

It is not clear yet how much we can do this year, especially since we will have to work with a number of other committees that share jurisdiction over personnel issues. So we have a long road to hoe. But that is only an even better reason to get started.

I look forward to working with your Subcommittee, Mr. Chairman, our full Committee, and with other committees, with NASA, and with the unions, and with all other interested parties to prevent this very foreseeable future from becoming a crisis. And let me report to the Committee some good news. The draft report on a new Department of Homeland Security was made available at 8 this morning, and I am pleased to report that it includes our provisions for an Undersecretary for Science and Technology. That is very good news. Thank you very much, Mr. Chairman, and with that, Mr. Administrator, I take leave to go about other very important business. Thank you.

[The prepared statement of Chairman Boehlert follows:]

PREPARED STATEMENT OF CHAIRMAN SHERWOOD BOEHLERT

Mr. Chairman:

Unfortunately, I can't stay at this hearing because I have to participate in the Intelligence Committee's continuing hearings on 9/11, but I did want to let everyone know some of my thoughts on the issues before you today. Also, I'm always pleased to see Administrator O'Keefe, and I wanted to ensure that he, in particular, understood where we are on these issues.

Like Administrator O'Keefe, I believe that there are few things more important for NASA than ensuring that it has the best possible workforce in future years. With the expected retirement bulge and the competitive market for top scientists and engineers, NASA may be left without the workforce it needs in the coming years. That's a situation we must avoid. NASA has to have the "cream of the crop" as it did at its inception.

So, one thing I want to see this committee do this year is to move forward with some proposals that would ensure that NASA has the people it needs. NASA has brought forward its own package, which we are reviewing, and I will carefully study the testimony you hear today. I want to understand better, for example, concerns raised by the unions.

But I do believe we can come up proposals that will help NASA and gain widespread support. It's not clear yet how much we can do this year, especially since we'll have to work with a number of other committees that share jurisdiction over personnel issues. So we have a long row to hoe. But that's only an even better reason to get started.

I look forward to working with our committee, with other committees, with NASA, with the unions and with all other interested parties to prevent this very foreseeable problem from becoming a crisis. Thank you.

Chairman ROHRABACHER. I am sure it is. Reading that full report has got to probably be about that thick. Thank you very much, Mr. Chairman, and we appreciate your attention to the Subcommittee but also the good work you are doing for the full Committee.

CHAIRMAN ROHRABACHER'S OPENING STATEMENT

Mr. O'Keefe assumed the role of NASA Administrator seven months ago, so after that six-month time period I am sure he decided what his priorities were going to be, and decided what direction he was going to go in. We are happy now, on the seventh month, to be with him to talk with him about the challenges he faces, challenges in restoring credibility to what I consider to be a neglected agency, neglected by some of the top leaders of the past Administration.

And I might add I think it is somewhat neglected by the current Administration as well in terms of the White House, and maybe that is something we should talk about. But we know that the Agency has been plagued by low morale, and it has been plagued by overdue decisions. These decisions were overdue perhaps because, as I say, the Agency has been neglected by the White House for a number of years.

I am pleased that Mr. O'Keefe is making progress in getting NASA back on track. As with any other endeavor, NASA's success hinges on its ability to attract a talented workforce, as Chairman Boehlert just mentioned, and I am sure we will hear more about workforce challenges today. I am also anxious to hear how Administrator O'Keefe plans to reaffirm the Agency's can-do spirit, so it can continue to inspire our hopes and our dreams for man's presence and utilization of space.

It is critical that NASA be an agency of innovators and of bold thinkers and of hard workers. Mr. O'Keefe has his challenges ahead of him, and we are anxious for him to succeed. I believe the answers to many of NASA's problems are simply good policies and good people. Mr. O'Keefe, we are here to listen, and we are here to help, but we must not forget that it will also take a concerted effort on the part of Congress and of the Bush Administration to insure that NASA remains what it has been called, "the jewel of the Federal Government."

We have assembled a panel of expert witnesses to provide us with their insight and to the critical issues that will be discussed today. NASA still has a long way to go before it is an affordable, well-managed science driven agency. And I believe today's hearing will help us better understand where NASA is, where it is headed, and how NASA's management reforms will prepare the Agency for the future.

[The prepared statement of Chairman Rohrabacher follows:]

PREPARED STATEMENT OF CHAIRMAN DANA ROHRABACHER

Today's hearing will examine the Administrator's vision for NASA, and NASA's legislative proposals for what is necessary to realize this vision. When Mr. O'Keefe assumed the role as NASA Administrator seven months ago, he faced the challenge of restoring credibility to a neglected agency plagued by low morale and overdue decisions. I'm pleased that he is making progress in getting NASA back on track. As with any other endeavor, NASA's success hinges upon its ability to attract a talented workforce, which I'm sure we'll hear about today.

I am anxious to hear how Administrator O'Keefe plans to reaffirm the Agency's "can-do" spirit so it can continue to inspire our hopes and dreams for man's presence in space. It is critical that NASA be an agency of innovators, bold thinkers, and hard workers.

Mr. O'Keefe has his challenge. We are anxious for him to succeed. I believe the answers to many of NASA's problems are simply good policies and good people. Mr. O'Keefe, we are here to listen and to help. We must not forget that it will also take a concerted effort on the part of Congress and the Bush Administration to ensure NASA remains the "jewel" of the Federal Government. We have assembled a panel of expert witnesses to provide us with their insights into these critical issues.

NASA still has a long way to go before it is an affordable, well-managed, science-driven agency. I believe today's hearing will help us better understand where NASA is, where it is headed, and how NASA's management reforms will prepare the Agency for the future.

I would now like to recognize Minority Ranking Member Bart Gordon from Tennessee for his opening statement.

RANKING MEMBER GORDON'S OPENING STATEMENT

Mr. GORDON. Thank you, Mr. Chairman, and welcome to our witnesses today. The main topic of today's hearing, NASA's workforce, is an important one. Without a strong and competent workforce, NASA cannot hope to carry out its mission in an efficient and effective manner. This morning we will hear NASA Administrator O'Keefe present his views on the workforce challenges confronting NASA, as well as the Administration's proposals for changing the laws that govern the NASA workforce.

We will also hear from a representative of one of the unions that represents a significant number of NASA employees. Those views of the likely effects of the Administration's proposals are in sharp contrast to the views expressed by the NASA Administrator. I think this will be useful for the Committee to hear these competing views. In some ways it echoes the current debate over how Federal workers should be treated in the proposed Department of Homeland Security.

Of course, this committee does not have jurisdiction over civil service workforce issues, and is not an area where we have much experience. Workforce issues are the responsibility of the Government Reform Committee. And so I hope this committee will not rush into taking positions on specific legislative positions until and unless we understand all of the implications of these provisions. At the same time, I think this committee does have a contribution to make in our oversight.

I mentioned earlier that without a strong and competent workforce, NASA cannot hope to achieve its goals. The converse is also true. Without clear and compelling goals and the funding to carry them out, the best workforce in the world will not be able to perform up to its potential. The NASA Administrator has sketched out his vision for NASA in his testimony.

However, this committee needs to know how that high level vision will be translated into specific programs and priorities. That is, what specifically does the Administration want NASA's workforce to be doing over the next five to 10 years, and is the Administration prepared to commit the funding necessary to make those programs credible. Only when we have this information can this committee effectively judge the appropriateness of NASA's workforce restructuring plans.

When NASA Administrator testified before this committee on February 27 of this year, I asked him for the results of NASA's strategic resource review. NASA had originally promised that SRR results would be released with the fiscal year 2003 budget request. We were also told that the SRR recommendations would be used to determine the future roles of NASA centers, what work would be contracted out, what workforce skills would be needed, how many employees would be needed, and so forth.

At the February 27 hearing, the Administrator acknowledged that the report was delayed but that he hoped to release the SRR report recommendations within the next few months. Since it is now mid-July, I hope you will be able to share the SRR recommendations with us today because they go to the heart of what the Administration intends to do with the NASA workforce. Without being able to examine the SRR recommendations, I don't know how this committee can properly assess NASA's proposed human capital provisions.

In addition, it is clear that the legislative provisions proposed by NASA only provide tools for restructuring NASA's workforce. They don't tell us how the Administration intends to use those tools. For example, the permanent buyout provisions and the industry exchange provisions, among others, seem to be geared to continuing downsizing of NASA's workforce, and transfer of more of its activities to the private sector. If this is true, then I would like to know how much more downsizing of NASA the Administrator is seeking, how quickly he intends to move NASA employees off the Federal payroll, and how he will insure that essential governmental responsibilities continue to be met.

The June 24 audit report of NASA Inspector General provides an important reminder that NASA cannot relax its attention to those government responsibilities. The audit report concerning NASA's oversight of the shuttle contractor safety procedures at the Kennedy Space Center found the following, and I quote, "The space flight operations contract states that NASA is to provide direct safety oversight of all U.S.A. operations. Nevertheless, Kennedy did not provide direct safety oversight of U.S.A.'s ground operations. Further, Kennedy did not perform any level of safety oversight for the integrated logistics and the high risk area of injuries and mishaps."

The audit went on to conclude implementing a level of oversight that contradicts that required by the space flight operations contract could lead to lapses in safety oversight, increasing the risk of harming personnel and damaging space shuttle hardware. Well I do not mean to belabor that point, but we can all agree that NASA is facing workforce challenges. What this committee needs to understand is whether the Administration's proposed legislative remedies, what they are, and whether they are going to meet these challenges. Again, I want to welcome the witnesses, and I look forward to your testimony.

Chairman ROHRABACHER. Thank you very much for some of those very provocative points. Mr. Lampson, I understand is going to pass on having a one-minute opening statement. But, Ms. Jackson Lee, do you have a very brief opening statement for us?

MS. JACKSON LEE'S OPENING STATEMENT

Ms. JACKSON LEE. Good morning. Thank you, Mr. Chairman. This is a very important hearing, and I am delighted to have the opportunity, Mr. O'Keefe, to hear from you this morning. Let me just at the onset indicate that a morning news show showed our astronauts interacting with youngsters at two museums, one in Houston and one in New York, emphasize the importance of developing new talent.

I am concerned that as we move forward we are extinguishing or denying the existing talent that we have and not finding alternatives to provide them the opportunity to serve their country. Congress is being asked to give NASA the flexibility to reshape its workforce, and even more to move to an alternative personnel system of its choosing on an expedited basis.

As that happens, I believe we have great concerns, and that is because NASA is becoming smaller and smaller, and older and older each year with no sign of rejuvenation in sight, and such a trend could eventually render NASA an exhibit in museums and history books instead of making it the leader in technology that we would like for it to be. I have a more complete statement but I believe that what we have is a great concern.

For example, just as an aside, I am working in my community on a scholarship science and technology program that I again will ask NASA to help and participate in, and I hope I will hear from NASA in terms of collaborating with my office on that and trying to encourage young people to participate in science and technology. But the key as I close simply is, and I ask unanimous consent to have the entire statement put in the record, we have decisions that are being made that I think impact negatively on the mission of NASA.

We have seen reports that suggest that we would go to Pluto and other places and yet the mission was eliminated from the 2003 budget. And finally we likewise had the Administrator cancel the X–38/CRV program. All these programs are valuable. All these programs need personnel. All these programs have valuable missions. And so I would be eager to hear how the Administrator plans to keep the mission, the vision, and the talented people that we have at NASA and grow more so that this nation can thrive in this world as we see it today. I yield back.

[The prepared statement of Ms. Lee follows:]

PREPARED STATEMENT OF REPRESENTATIVE SHEILA JACKSON LEE

Mr. Chairman,

Thank you for organizing this important and timely hearing to discuss the workforce and management challenges that face NASA.

I think every member of this committee and most of my constituents back home in Texas know my passion for NASA and all of the exciting work they do. NASA plays many roles, and means so much to America today. NASA is a source of dreams for our young and old alike. It provides insights into the origins and destiny, and wonder, of our universe. On the way to this noble goal, NASA develops innovations that spur on our economy and keep us on the cutting edge of technology.

NASA inspires young engineers and scientists to push their minds to new levels of excellence. These people become role models for future generations of intellectual pioneers. Just last week we in the Committee on Science passed unanimously an amendment stating the sense of Congress that the future Secretary of Homeland Security should tap the fabulous knowledge and resources at NASA when designing barriers to cyber terrorism. Distinguished guests, I feel that you at NASA are in integral part of the future of the United States.

That is exactly why I am deeply troubled by the direction this important program in taking. NASA is becoming smaller and smaller, and older and older each year, with no sign of rejuvenation in sight. Such a trend could eventually render NASA an exhibit in museums and history books, instead of making it the leader in technology and exploration that it should be. You have been plagued by accounting and cost-overrun difficulties, which threaten long-term support, and jeopardize critical projects.

As I mentioned before, NASA is a source of dreams and role models for our future scientists, engineers, and all sorts of people who want to a part of something truly great and noble. However, I continually get mail from the Houston area expressing concern over the dearth of senior level minorities and women in the NASA administration. I understand that getting qualified people through the ranks and into high-level positions takes time, but there should be some long-term strategy to address this issue, and some long-term progress.

Furthermore, at NASA recently, there seems to be a fundamental disconnect between logic and policy. It seems that scientists would appreciate good studies. But,

NASA is consistently commissioning studies and reviews, only to ignore their recommendations. For example, a National Academy of Science review panel commissioned by NASA ranked a mission to Pluto, the Kuiper Belt, and Europa as being of the highest scientific priority—yet this mission was eliminated from the FY 2003 budget request.

The NASA Administrator has stated that he will only fund a reduced-crew, "Core Complete" Space Station, although an independent task force, again commissioned by NASA, concluded that a such a slimmed-down force would not even be capable of carrying out the "high priority" list of research projects on the station.

And most frightening for our young astronauts, although the Administration has been repeatedly urged to develop a U.S. Crew Return Vehicle (CRV) as soon as practicable, the NASA Administrator has canceled the X–38/CRV program without identifying a viable alternative for rescuing crew in the event of an emergency.

Where is NASA going? I am looking for some thoughtful, creative strategies to get NASA back on track, not a series of half-baked stop-gap measures to satisfy political whims. Today we are focussing on workforce issues. Of course, getting qualified people to come and stay at NASA is a critical component of such a long-term strategy to get NASA doing the work it needs to do. But I am reluctant to give broad latitude to an Administrator to develop a workforce, *without any clear vision of what that workforce will be intended to do.*

Many of the changes the Administration is requesting would produce inequalities between NASA employees and other branches of the Federal Government, that may create brain-drains at other agencies. The proposed changes are probably outside the jurisdiction of this committee anyway.

We do not want to create an imbalance or unfairness in our federal workforce, and potentially cripple it, in the name of a poorly defined vision at NASA. We should not allow the Administrator to squander already scare funds in the name of securing qualified professionals, if the work of these professionals will be outsourced to the private sector in six months.

We are still awaiting a coherent vision of NASA's future. Many of NASA's major programs are under review. The results of the Strategic Resources Review are still a mystery.

I look forward to discussing today the policy changes the Administration is proposing to address workforce issues at NASA, to determine if any are appropriate or have a chance of producing meaningful improvement. However, I do not want to help the Administration run faster, if I am not convinced that they are running in the right direction.

This hearing was meant to address "Workforce *and Management* Challenges at NASA." Perhaps we should dedicate a bit more energy to the "Management" side of that task.

Thank you.

Chairman ROHRABACHER. Thank you very much, Ms. Jackson Lee. The opening statements of other members will be put in the written record so we can get right to the testimony today. And hearing no objection as to that, so ordered.

[The prepared statement of Mr. Etheridge follows:]

PREPARED STATEMENT OF REPRESENTATIVE BOB ETHERIDGE

Thank you Mr. Chairman for having this hearing. I also wish to thank the distinguished panel for their testimony.

I am concerned about the many challenges that NASA is facing today and appreciate the effort which NASA is making to address many of them. As former Superintendent of Schools in my state of North Carolina, I have a great interest in education policy. Therefore, I am especially concerned in the challenges that NASA is facing in terms of its workforce. NASA, and the Federal Government in general, are confronted with the problem of attracting and retaining a skilled workforce. What is being done about NASA's shrinking workforce of scientists and engineers? What is being done to inspire the next generation of scientists so that we can maintain a highly skilled science and engineering workforce in the future? Our pool of college students enrolled in science and engineering courses is declining. NASA, more than any other agency, relies on a highly skilled scientific and engineering workforce to accomplish its mission. The workforce and management challenges facing NASA are clear. What is not clear to me are the major steps being taken to chart a clear course for the future. NASA's current workforce is approaching retirement eligibility. What is being done to address the workforce needs of the future? How can Congress help NASA meet these challenges?

NASA has responsibilities that are widespread and like no other agency. NASA's mission encompasses human exploration and development of space, the advancement and communication of scientific knowledge, and research and development of aeronautics and space technologies. Its activities span a broad range of complex and technical endeavors—from investigating the composition, evaluation, and resources of Mars; to working with a international partners to complete and operate the International Space Station; to providing satellite and aircraft observations of earth for scientific and weather forecasting purposes; to developing new technologies designed to improve air flight safety. What is the biggest management challenge facing NASA? What can Congress do to help NASA meet this challenge?

Chairman ROHRABACHER. The Chairman also requests unanimous consent for the authority to recess this hearing at any point. Hearing no objection, so ordered. I also ask unanimous consent to insert at the appropriate place in the record a background memorandum prepared by the Majority for this hearing. And hearing no objection, so ordered.

HEARING CHARTER

SUBCOMMITTEE ON SPACE AND AERONAUTICS COMMITTEE ON SCIENCE U.S. HOUSE OF REPRESENTATIVES

NASA Workforce and Management Challenges

THURSDAY, JULY 18, 2002
10:00 A.M.–12:00 P.M.
2318 RAYBURN HOUSE OFFICE BUILDING

1. Purpose

On Thursday July 18, 2002, at 10:00 a.m. in room 2318 Rayburn, the Subcommittee on Space and Aeronautics will hold a hearing on National Aeronautics and Space Administration (NASA) Workforce and Management Challenges. The hearing will examine the Administration's goals for transforming the Agency over the next five years, the top management challenges that must be overcome to realize these goals, the challenges NASA faces in reshaping and restructuring its workforce, the need for legislation to provide new and expanded authorities for recruiting and retaining a high-quality workforce, and how NASA's legislative proposals on human capital support the transformation of the Agency and its workforce.

The witnesses for this hearing are:

- The Honorable David Walker, Comptroller General of the United States
- The Honorable Sean O'Keefe, NASA Administrator
- Mr. Mark Roth, General Counsel of the American Federation of Government Employees

2. Major Issues

- **Administration's Goals for Management Reform:** In a speech last April, the NASA Administrator articulated a new vision and mission for the Agency, as well as how NASA would carry out the Administration's management agenda through its Freedom To Manage initiative. Mr. O'Keefe stated that NASA's management is to be 1) Citizen-centered; 2) Results-oriented; 3) Market-based, actively promoting innovation through competition; and 4) Focused on that which only NASA can do and to avoid duplicating efforts. Major Questions: What are the challenges to implement these reforms? How will these reforms change NASA as an agency and how will they effect NASA's programs?
- **General Accounting Office (GAO) Reports on NASA's Major Management Challenges and Program Risks:** The GAO identified significant management risks in 1) financial and contract management; 2) cost control on the International Space Station; and 3) recruiting and retaining a high-quality workforce. Specifically, the GAO identified the need for NASA to better integrate human capital plans into its workforce management strategies. Major Questions: What action has NASA taken in response to GAO's recommendations? What additional action does NASA need to take to address GAO's concerns?
- **NASA's Workforce Changes:** NASA relies on a highly skilled aerospace scientific and engineering (S&E) workforce to accomplish its mission. However, this workforce is suffering from a lack of young scientists and engineers in NASA's workforce pipeline. For example, NASA's scientists and engineers age 60+ years old (Apollo era) outnumber the under-30 workforce by 3 to 1, and approximately 25 percent of NASA's S&E workforce is eligible to retire within the next five years. NASA down-sized from 25,000 civil servants in 1993 to just under 18,800 civil servants in 2002 (a decrease of 30 percent). This resulted in skill gaps for certain NASA programs and technical areas, like systems engineering, propulsion, and robotics. Major question: How can NASA reshape and revitalize its workforce to meet the challenges of NASA's future missions?
- **NASA's Legislative Proposals for Human Capital:** NASA has proposed new and expanded legislative authorities to provide the Agency additional flexibility

to recruit, hire, and retain critical skills. In addition, NASA has proposed a scholarship-for-service bill which would provide scholarships to students studying NASA-related disciplines, such as science and engineering, in exchange for a service obligation to the Agency. NASA has also requested authority to offer voluntary separation and early retirement incentives to help reshape the current workforce. Each legislative proposal is briefly described in Attachment A. Major Questions: What specific workforce problem does each of the proposed legislative provisions solve? Has NASA fully exploited its current civil service authorities before it asks for additional flexibilities? How do NASA's legislative proposals contrast with existing civil workforce authorities?

3. Workforce and Management Challenges

GAO, the NASA Inspector General, the NASA Advisory Council, and the Congress have expressed grave concerns about NASA's poor management of its space programs. These concerns have been primarily due to a lack of credible accounting practices, poor financial and contract management, and the inability to control cost growth on major programs. Even beyond the technical and scientific challenges, NASA's greatest long-term management challenge is its ability to attract and retain the skilled workforce needed to accomplish its mission.

NASA Administrator O'Keefe said during testimony last April before the National Commission on the Future of the Aerospace Industry that he has taken a fresh look at the long-term management, resource, and technical challenges facing NASA. To develop these management reforms, NASA is primarily using the five major initiatives from the President's Management Agenda (human capital, competitive sourcing, improved financial management, e-government, and integrating budget and performance) as a guide to enact management reforms within the Agency. Some of the reforms include human capital strategic planning; competitively sourcing Space Shuttle operations; accelerating standardized accounting and finance software across the Agency; and transitioning to full cost accounting in its program budgeting.

In order to enact human capital reforms, NASA has proposed legislation to modify existing laws governing its civil service workforce. The overarching reason for these legislative proposals is to correct scientific and engineering (S&E) skill imbalances in NASA's technical aerospace workforce. The skill imbalances that NASA needs to correct are primarily due to a decline in the overall number of aerospace scientists and engineers in the United States as well as NASA's civil service downsizing during the 1990s. NASA points to six national trends adversely affecting its ability to attract and retain a highly skilled S&E workforce: 1) Shrinking Scientist & Engineer (S&E) Pipeline; 2) Increased Competition for Technical Skills; 3) Lack of Diversity in Applicants; 4) Skills Imbalances and Lack of Depth; 5) Knowledge Loss due to Retirements; and 6) Greater Recruitment/Retention Problems for S&Es. NASA's legislative proposals seek greater flexibility in civil service laws in order to respond to these national aerospace S&E workforce trends. However, as noted earlier, NASA's management challenges are not simply technical in nature and may require other skills, such as business management, accounting, and cost estimating.

4. National Trends in the Aerospace Scientific and Engineering Workforce

In testimony before the Space & Aeronautics Subcommittee on May 15, 2001, General (USAF, Ret.) Tom Moorman, who led an aerospace industrial base study, testified: "Our discussions with the leadership of twenty-one space companies underscore the fact that human resource issues are the largest long-term problem the industry faces. Even as companies continue to consolidate, the stock of human capital has been declining sharply." He explained that a declining number of new college recruits educated in aerospace science and engineering as well as experienced middle level managers are resulting in far greater technical and management risks in aerospace programs. He testified: "If present trends continue, the space industry may lack the critical mass of talent required to design and integrate complex spacecraft."

Several studies provide evidence for this conclusion. Over half of the current workforce is over 45 years old, and the mean age for aerospace engineers is eight years higher than for scientists and engineers in other technical fields. According to industry human resource managers, the supply of aerospace workers between ages 35 and 44 years is not sufficient to meet the industry demand due to the departure of many mid-level managers to other industries like telecommunications or information technology. Likewise, the pipeline of people entering the aerospace industry is shrinking as graduate enrollment in aerospace engineering declined 16 percent from 4,036 in 1992 to 3,407 in 2000. These conditions in the aerospace industry and a shrinking pipeline of new aerospace scientific and engineering talent have a direct impact on NASA's ability to recruit and retain technical talent.

5. NASA's Workforce

Successful scientific and engineering (S&E) research and development organizations must provide for a steady infusion of new technical talent to ensure organizational health and vitality, and to ensure that the organization is prepared for future missions. NASA's science and engineering workforce is one of the most technical and highly-trained in the Federal Government. Over 50 percent of NASA's S&E workforce has a graduate degree (Masters or Ph.D.). However, NASA's ability to attract and retain new talent is threatened by increased competition from other industries.

Government scientists and engineers typically make less in salary and bonuses than their industry or academia counterparts. NASA recruiters state that top candidates just out of college were receiving offers $10–$20,000 more than the entry-level salaries they could offer. Similarly, recruiting, retention, and relocation bonuses for NASA are less than industry standards. However, NASA may not be adequately using its current human resources authorities to attract and retain a highly-skilled NASA workforce. For example, NASA reported that only 38 retention bonuses were offered to NASA workers last year and only 18 percent of new hires were offered recruitment bonuses.

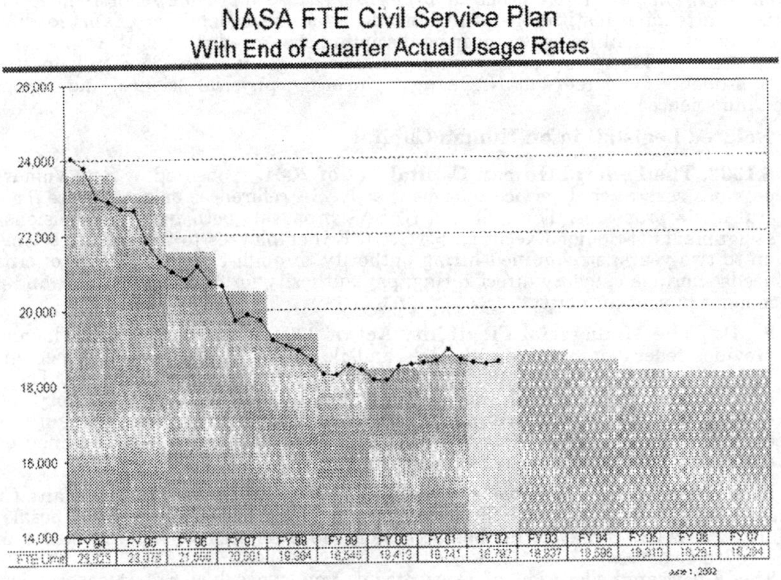

NASA's workforce has changed significantly over the past ten years, due primarily to voluntary separation incentives (buyouts) and early retirement (early-outs). More than 3,500 employees left NASA voluntarily during three buyouts offered in the mid-1990s, and overall, NASA downsized from 25,000 civil servants in 1993 to approximately 18,800 civil servants in 2002 (an almost 30 percent decrease). The chart above shows the declining trend in the number of NASA's Full-Time Employees (FTE). NASA's current FTE ceiling is 18,792, but only 18,450 positions are actually filled. Since the beginning of fiscal year 2002, NASA has hired 666 individuals and lost 627 workers through attrition.

NASA now recognizes that the downsizing during the 1990s created skill imbalances and gaps with its government workforce, because there was little control over which employees departed the government and little ability to ensure that the right balance of skills remained within the government. Thus, NASA's current problems are partially a result of poor human resource management in the past. This skill imbalance may affect NASA's ability to perform future missions and manage its programs. NASA has reported problems in recruiting new employees to work on the Space Shuttle and International Space 9 Station programs.

6. NASA's Human Capital Legislative Proposals: Reshaping the Aerospace S&E Workforce

In order to begin to address these human capital challenges, NASA initiated several programs, such as the National Recruitment Initiative and Strategic Human Capital Implementation Plans, with the aim to reshape and reconstitute the NASA workforce with the right skill mix. Some of these initiatives have resulted in changes to NASA's internal human resources policies. NASA also requested that Congress consider the legislative proposals summarized in Attachment A to provide the Agency greater flexibility than is allowed under current regulations and statutes. Attachment A is NASA's description of each human capital proposal along with the Agency's supporting rationale and related initiatives/precedents already enacted by other federal agencies or in other civil service legislative proposals before Congress.

Certain legislative proposals, such as the scholarship program and enhanced retention bonuses, are tailored to meet NASA's need to recruit from a shrinking number of aerospace scientists and engineers, while other proposals such as streamlined hiring authority are aimed at NASA's recruiting problems. Some proposals are tailored to attracting executive-level industry and academic expertise into the Agency through funding incentives and exchange programs.

Scholarship for Service: NASA Science and Technology Career Enhancement Act: This legislative proposal authorizes NASA to provide scholarships to U.S. citizens attending institutions of higher education in academic programs leading to a degree in a scientific or engineering discipline area needed by NASA, in exchange for service to the Agency after graduation. The program is designed to help attract U.S. students for careers at NASA in engineering, physical science, and biological and life sciences.

7. Related Legislation on Human Capital

- **S.1603, The Federal Human Capital Act of 2001,** sponsored by Sen. Voinovich, provides several civil service reforms that NASA references as precedence for several of its proposals. This bill and NASA's proposals both include provisions for assignments of Senior Executive Service (SES) employees to the private sector for up to two years; streamlined hiring authority through category ratings of critical needs/shortage category direct hiring; pay authority for critical positions; and permanent buyout authority.
- **S.1612, The Managerial Flexibility Act of 2001,** sponsored by Sen. Thompson, provides federal managers with tools and flexibility in areas such as personnel, budgeting, property management and disposal, and for other purposes. This bill and NASA's proposals both include streamlined hiring through a category rating system; provide larger recruitment, relocation, and retention bonuses; authorizes early retirement for workforce restructuring; and authorizes a streamlined demonstration and alternative personnel system.

Both pieces of legislation were referred to the Senate Governmental Affairs Committee. The Federal Services Subcommittee held a hearing on both proposals on March 19, 2002. However, further action on the legislation is not expected during the 107th Congress.

The American Federation of Government Employees has raised several issues with these legislative proposals as part of the Administration's government-wide proposals. The AFGE expressed concerns that these initiatives will enrich top management but not rank and file workers and that the funds for recruitment, retention, and relocation bonuses would come out of other worker's regular salary accounts. The AFGE claims that some proposals would increase the likelihood of federal workers losing their jobs to contractors and losing their pay raises in order to pay for the bonuses of others. Further, the AFGE expressed concern that the current "human capital crisis" was self-inflicted and a direct result of the Federal Government downsizing of 375,000 federal jobs.

8. Witnesses

The Honorable David Walker, U.S. Comptroller General, has been asked to address the following:

- What is the GAO's perspective on NASA's top management challenges? Specifically, address NASA's challenges in human capital as well as financial and contract management.
- What recommendations would you give NASA to meet these challenges, especially in the area of managing its human capital?

- How do the human capital challenges facing NASA contrast with the entire Federal Government?
- To what extent are current civil service authorities not sufficient to address the Administration's human capital management challenge?
- Are NASA's legislative proposals sufficient to transform its workforce?

The Honorable Sean O'Keefe, NASA Administrator, has been asked to address the following:

- What are the management goals for NASA over the next five years? Specifically, address the goals for financial management and workforce transformation.
- NASA initiated a series of studies this past year regarding NASA's core competencies (Strategic Resources Review), Space Shuttle competitive sourcing, and workforce adjustments to the Space Station program. Are NASA's workforce transformation goals supported by these studies?
- How do NASA's legislative proposals on human capital support the transformation of the Agency and its workforce?

Mr. Mark Roth, General Counsel for American Federation of Government Employees, has been asked to address the following:

- What do you consider to be the most important challenges facing the NASA workforce over the next five years?
- What needs to be done to ensure that NASA is able to attract and retain a highly capable workforce to meet the Agency's future needs?
- If enacted, do you think NASA's proposed human capital legislative provisions would provide the Agency the necessary tools to successfully transform its workforce?
- Are there any alternative legislative provisions that AFGE would recommend?

ATTACHMENT A. NASA's Human Capital Legislative Proposals (Prepared by NASA)

PROPOSAL	DESCRIPTION	RATIONALE	RELATED INITIATIVES/PRECEDENTS
Knowledge Management and Pipeline			
NASA Science and Technology Career Enhancement Act of 2001--Submitted to Congress on July 20, 2001	Allows NASA to provide scholarships to U.S. citizens attending institutions of higher education in academic programs leading to a degree in a discipline area needed by NASA, in exchange for a service to NASA requirement.	Over the next two decades, the number of undergraduate and graduate students projected to enter the science and engineering job pipeline may not be sufficient to meet the agency's needs. To address this shortfall, the NASA Science and Technology Scholarship Program is designed to help guide U.S. students toward careers in engineering, physical science, and biological and life sciences. In return for scholarships, students will be required to fulfill a year-for-year service requirement at NASA following graduation. NASA anticipates, subject to available funding, a sustained program level of approximately 300 students.	NIH Undergraduate Scholarship Program; provides service-obligated scholarships to students in course of study for professions needed by NIH. National Health Service Corps Scholarship Program of HHS Scholarship for Service Program of NSF.
Extension of Inter-governmental Personnel Act (IPA)[1] Period	Permits IPA assignments for up to 6 years for long-term projects: limited to 10 individuals per year. Amends Title 5 to apply this provision Governmentwide.	IPA assignments are limited to 4 years by statute. Although the need for such assignments beyond 4 years would be infrequent, the nature of program and project work at NASA is such that assignments of longer duration may be desirable to avoid undue disruption to a program/project caused by terminating a critical IPA assignment.	

[1] *(IPA) Intergovernmental Personnel Act is a current authority for service in the Federal Government that provides for the temporary assignment of personnel between the Federal Government and state and local governments, colleges and universities, Indian tribal governments, federally funded research and development centers, and other eligible organizations.

NASA-Industry Exchange Program	Establishes IPA-like program to permit temporary assignment of personnel from "for profit" industries to NASA and vice-versa.	The IPA program provides a valuable vehicle to enable the exchange of ideas and expertise between NASA and academia, but there is no comparable mechanism for use with industry. Through assignments of NASA personnel to industry and vice versa, the Agency would gain a valuable source of knowledge – technical, managerial, and entrepreneurial – to enhance its core competencies and improve business practices during a time of competitive sourcing and transformation. Participants would be subject to conflict of interest and ethics statutes.	The Administration's Proposal to the DoD FY03 Auth Bill: Establishes pilot program for temporary assignment of employees of private sector commercial organizations to non-management acquisition or technology positions in DoD at levels equivalent to GS-11-15. Also establishes pilot for assignment of private sector employees in acquisition field to DoD positions. Both provisions amend title 10, and would apply to DoD only. H.R. 3925 (Digital Tech Corps Act): Creates exchange program allowing Federal agencies and private sector companies to exchange mid-level IT managers for 1-year assignments with optional 1-year extension. S.1603 (Voinovich bill): Provides for assignments of Senior Executive Service (SES) employees to the private sector for up to 2 years, modeled on IPA program.
Recruitment and Retention			
Recruitment, Relocation, Retention Bonuses (3 Rs)	Authorizes larger bonuses, based upon the length of service period, capped at 25% of *adjusted* salary X number of years of service agreement, up to 4 years; provides flexible payment provisions (lump sum or installments); permits payment of retention bonus to individual who would otherwise leave for a different agency.	Current authority is not sufficiently generous (recruitment & relocation bonuses limited to 25% of base salary for 1 year) or flexible (only 1 method of payment; retention allowances not allowed if employee has *Federal* job offer) to hire or retain highly qualified individuals. Changes would address expected attrition rates and loss of institutional knowledge; and compete with the private sector for the very best candidates.	S.1612: Same provisions IRS Reform Legislation: Provides for variations from regulations governing the 3R's within IRS, for a period of 10 years. S.1639: Essentially the same as S.1612.

Streamlined Hiring	(1) Authorizes direct hiring for positions in "critical needs" or "severe shortage" categories and (2) Authorizes category rating system for evaluating candidates (for any position).	Cumbersome, lengthy application and rating processes impede NASA's ability to compete with the private sector for critical hires, especially in the science and engineering (S&E) fields. As the S&E pipeline shrinks, a streamlined hiring process is crucial to allow NASA to address competency gaps in implementing SRR actions. Based on an OPM-approved plan, NASA would be able to eliminate the lengthy evaluation process where there is an identified shortage of qualified candidates or a "critical need." Category rating provides a more flexible, streamlined approach to evaluating candidates for any position, without the need to assign numerical points to each applicant. Job offers can be made more quickly, and managers have the option to select from among candidates in the top quality group, rather than being limited to the top 3 candidates. Veterans are placed at the top of the appropriate category.	S.1612: Same provisions S.1603: Provides for category ratings and critical needs/shortage category direct hiring, subject to approval of implementing plan by OPM. The Administration's Proposal to the DoD FY03 Auth Bill: Provides for direct hire authority for shortage and critical need hiring; veterans preference applies. Authority limited to DoD. DoD bill has same category rating provision, proposed Federal-wide. USDA Alternative Personnel System: Uses category rating system. Demo Projects: Numerous demos use category rating system. IRS Reform Legislation (P.L. 105-206) Provides for category ratings. S.1639: Provides for category ratings.

Conversion of term employees to permanent and allow appointments up to 6 years	(1) Authorizes non-competitive conversion of a term employee to a permanent position provided the original competitive announcement indicated the possibility of such conversion and the position is in the same line of work. If these conditions are not met, a merit promotion competition would be required for conversion to permanent employment. (2) Authorizes term appointments for up to 6 years for project-type work; limited to 10 per year without OPM approval. Amends Title 5 to apply this provision Governmentwide.	Current rules require a second, nationwide competition to convert a term employee to permanent civil service status. This puts an unreasonable burden on agencies and creates a "sham" competition when no actual vacancy exists. Under the new authority, NASA could convert a term employee to permanent after at least 2 years of employment, provided the possibility was made known in the original public notice, and the position is essentially the same. Much of NASA's work is project oriented. Research and development projects may extend to 5 years or more. Current rules restrict term appointments to 4 years, with extensions possible only with OPM approval. Allowing 6-year appointments would make this authority more flexible and useful, more in line with NASA's core work. It would facilitate the influx of talent from the private sector and academia to work on projects or programs of limited duration, without creating a permanent employment situation.	The Administration's Proposal to the DoD FY03 Auth Bill: Same provision; limited to DoD (amends title 10). IRS Reform Legislation: Includes similar provision. Demo Projects: Several DoD demos have similar provision.
Increase the maximum rate of pay for NASA-Excepted employees (NEX)	Increases the maximum rate of pay for NEX employees to level III of the Executive Schedule	Current Space Act provisions limit maximum rate of pay for NEX employees to level IV ($130,000) of the Executive Schedule. This cap should be raised to level III ($138,200) to provide greater flexibility to the Administrator in hiring exceptional talent from academia and the private sector.	This provision applies to a NASA-specific hiring authority in the Space Act. There is no impact on other agencies.
Pay Authority for Critical Positions	Authorizes the Administrator to set pay up to level of Vice President for critical positions requiring expertise of extremely high level. Up to 10 employees may be paid under this authority at any time. Appointments into NASA under this provision are limited to 4-year terms. 10 year sunset provision.	Currently, requests to pay an individual under this authority must be approved by OMB in consultation with OPM, and monetary amount is limited to rate of level I of Executive Schedule. Exceptional individuals are needed to ensure NASA's success in implementing transformations in the coming years. In order to extend timely offers to such candidates, approval authority must be delegated to Administrator. Amount of pay should be increased to enhance NASA's ability to attract world-class talent.	IRS Reform Legislation: Same provisions, but IRS legislation provides that 40 critical pay positions may be filled. S.1603: Same provisions, but allows agency heads to establish no more than 1 position for every 3,000 employees of the agency.

Transformation			
Permanent and enhanced buyout authority	Provides permanent buyout authority to NASA; use is contingent upon OMB approval of plan. Provides for waiver of FTE reduction to allow for restructuring and permits targeting to be based on skills. Retains the monetary "cap" on buyouts of $25,000. Authorizes the Administrator to increase the buyout amount up to 50% of an employee's rate of pay for critical restructuring needs, based on criteria specified in Agency plan (limited to 10 per year without specific OMB approval).	Current NASA buyout authority expires Sept. 02. Expected workforce transformations (ISS, SSP, SRR) will require voluntary separation incentives to balance workforce and avoid RIF. While retaining the monetary cap on buyouts or $25,000 for most employees, allows limited flexibility to the Administrator to offer up to 50% of salary to up to 10 individuals per year (or more, with specific OMB approval). This provision addresses the challenges of the eroding value of a $25,000 buyout in today's economy, reflects consumer price index increases since the buyout amount was established in 1992, and provides a more effective incentive for voluntary attrition. Given the relatively high salaries among the NASA workforce, a $25,000 buyout may be inadequate to encourage needed attrition in some cases; the provision to increase the amount to 50% of rate of pay will address that challenge.	S.1612: Same provisions, but does not increase monetary amount. S.1603: Provides permanent buyout authority to agencies, contingent upon OMB approval of plan. Does not increase monetary amount. S.1639: Essentially the same as S.1612 by including waiver of FTE reduction and ability to target based on skills. Does not increase monetary amount. (Note: provisions of S.1612, S.1603, and S.1639 are governmentwide).
Expanded use of early retirement	Authorizes early retirement for workforce restructuring, not just downsizing. May be targeted based on same criteria as buyouts (e.g. skills).	To rebalance workforce skills post-downsizing, and for expected workforce transformations (ISS, SSP, SRR), need ability to offer early retirements without loss of FTE.	S.1612: Same provision. S.1639: Essentially the same as S.1612.

Demonstration Project Authority			
Streamlined Demonstration and Alternative Personnel System Authority	Authorizes NASA to request and implement a demonstration project, subject to OPM approval, following a streamlined process and without any limitation on the number of employees that would be covered by the project. Provides a mechanism for converting a successful demonstration project, upon approval by OPM, to a permanent Alternative Personnel System, without the need for special legislation	Current process for seeking demonstration authority to test innovative human resources flexibilities is burdensome and time-consuming; in addition, the authority limits project coverage to only 5,000 employees. Consequently, the potential benefit of this authority is not worth the time and effort required to design, gain approval, and implement demonstration projects. This proposal streamlines the process for obtaining approval of, and implementing, a demonstration project. In addtion, there would be no limit on the number of employees covered by the project. If successful, a demonstration project could become permanent (as an "Alternative Personnel System") through approval by OPM, rather than by legislation.	S.1612: Same provision. The Administration's Proposal to the DoD FY03 Auth Bill: Same provision; limited to DoD. S.1639: Same provision. IRS Reform Legislation: Similar provision.

Chairman ROHRABACHER. Unfortunately, that sounds like a vote. Lastly, let me ask unanimous consent for the record—oh, it is a recess. It is a recess—that this hearing remain open until July 25, 2002, so that additional testimony may be inserted into the record. Without objection, so ordered.

We have a distinguished panel, as I mentioned, with us today. We have an accountant, we have a lawyer, and we have a public administrator. Now there has to be a joke in there somewhere. I will spare everyone. They are here to give us their perspectives concerning NASA's workforce and management challenges. And we have asked them to summarize, and we would suggest, if you could, summarize your testimony to five minutes so that we can engage in some dialogue and some questions and answers.

Our first witness is David Walker, Comptroller General of the United States, and head of the General Accounting Office that supports the Congress and provides us with the professional nonpartisan reviews of government operations. Luckily for us, David is also a human capital expert, which is somewhat the focus of the hearing today. He has written several books and articles on the subject. Thank you very much, David, for joining us today, and you may proceed with your testimony.

STATEMENT OF HON. DAVID M. WALKER, COMPTROLLER GENERAL OF THE UNITED STATES, GENERAL ACCOUNTING OFFICE

Mr. WALKER. Mr. Chairman, Ranking Member Gordon, Members of the Committee, it is a pleasure to be here. Yes, I am a CPA, but I don't consider myself an accountant. I am an accountability professional. But with that, I have been asked to address some of the major management challenges and performance challenges that NASA faces based upon our last performance and accountability report that was issued in January 2001, that will be updated in January 2003.

I will focus my remarks, Mr. Chairman, on the human capital challenges although my statement includes information on all four of the challenges that we outlined in our report of January, 2001. Recognizing the need for change, NASA's new Administrator, Sean O'Keefe, has recently articulated a new vision for NASA, one that is science driven, not destination driven. To put NASA on a better footing to fulfill this mission, the Agency is taking on a major transformation effort.

TRANSFORMATION TAKES MANY YEARS

Importantly, no matter whether you are in the public sector, the private sector, or the not-for-profit sector, committed and sustained leadership from the top as well as modern, effective, credible, and equitable human capital or people strategies are the key to any successful transformation effort. Although NASA is in the very early stages of this transformation, it is already undertaking initiatives designed to reshape and strengthen its workforce, including developing a strategic human capital plan, an agency wide workforce planning and analysis system.

The Subcommittee is also considering certain legislative flexibilities that NASA is seeking in order to help transition and to help transform itself. As I mentioned, the human capital challenge is arguably the most important because that is the key to the future. That is the key to success for not only NASA but quite frankly any government agency. This will require concerted effort and sustained attention by NASA's top leadership in order to effectuate this change.

It will take many years. A vast majority of this can be done within the context of current law, but we do believe that NASA will need the Congress' help in order to provide reasonable flexibility with appropriate safeguards to prevent abuse of employees. We have been provided some of this flexibility, quite frankly, ourselves, Mr. Chairman, at the GAO. In addition to that, the Transportation Security Agency, as well as the IRS and others from time to time have been provided some types of flexibilities that the Congress and this committee can look to.

It is important to note that based on my past and present experience in the management area, which includes a number of cultural transformation efforts for large and complex organizations, that to make this work, to not only design the plan but to implement the plan and to make it stick, which is important, and to enable it to stick irrespective of who the Administrator is, irrespective of who the President is, irrespective of other key players, will take at least five to seven years guaranteed.

CONCEPT OF A CHIEF OPERATING OFFICER

That is what it takes in the private sector. That is what it is going to take in the public sector. Therefore, one of the things that I would commend to this committee is the concept of a Chief Operating Officer as possibly a way to assist Administrator O'Keefe to focus on the strategic planning, to focus on the implementation, to focus on achieving this cultural change that will take years to do, and to consider the possibility of such an individual that would be a term appointment with a performance contract to try to help in this effort.

I made the same comment last night before the Select Committee of this House with regard to the Department of Homeland Security, and hopefully they will consider it in that context.

NASA is just beginning to undertake major efforts to transform itself, but it is off to a strong start. Mr. O'Keefe came on board as the Agency's new Administrator in January, 2002. He began articulating his vision in April. Concurrently, he began undertaking a number of human capital and other management initiatives designed to foster better financial management, information technology management, and budget and performance integration.

In view of these efforts, just this week the Office of Management and Budget identified NASA as leading the government in progressing toward the implementation of five government-wide changes contained in the President's management agenda. I would note that only includes the Executive branch. While NASA's progress is noteworthy, it is currently rated as a red light on each of the five key areas, and so they still have major problems, major challenges, but they are making good progress on what will take a number of years to effectively achieve.

NASA'S HUMAN CAPITAL PLAN

NASA has also developed a strategic human capital plan that incorporates strategies, tactical actions, and metrics to support human capital goals. The plan has been submitted to OMB and OPM for review. The plan is based on the planning model that was developed by OPM and another one that was developed by GAO, which was published in March of 2002.

As I mentioned, the Subcommittee is considering certain legislative flexibilities. We believe that some flexibilities would make sense provided they have adequate safeguards to prevent abuse.

We also believe that modernizing the Agency's performance management system to have modern, effective, credible, and equitable performance appraisal and management systems will be key to a successful transformation effort. Those do not take a change in law. It is critically important, however, that they be done.

In closing, NASA has a long and proud history and does many things well. But times have changed and NASA must change with the times and considering what it does and how it does business.

Moreover, the Agency is facing significant management challenges, which if not effectively addressed stand to hurt NASA's credibility with the Congress and its partners and hamper important space and other technology and science missions. I would like to commend Mr. O'Keefe for his personal commitment to help transform NASA, and he is off to a fast start on what will be a long march. Thank you, Mr. Chairman.

[The prepared statement of Mr. Walker follows:]

PREPARED STATEMENT OF DAVID M. WALKER

NASA Management Challenges

Human Capital and Other Critical Areas Need to be Addressed

Mr. Chairman and Members of the Subcommittee:

I appreciate the opportunity to discuss the major management challenges and program risks facing the National Aeronautics and Space Administration (NASA). Right now, NASA is at a critical juncture. Clearly, since its inception, NASA has advanced space exploration and scientific knowledge and accomplished unparalled feats of engineering. But NASA is now facing difficulties, particularly in terms of maintaining a skilled workforce, controlling costs, and providing effective oversight for important projects. Such problems have been debilitating to important space missions. For example, substantial space station cost growth, which NASA became aware of in early 2001, has resulted in cutbacks in construction, the number of crew members, and scientific research, and, in turn, raised concerns about the viability of the program and has negatively impacted the Agency's credibility with the Congress.

Recognizing the need for change, NASA's new Administrator, Sean O'Keefe, has recently articulated a new vision for NASA—one that is science-driven, not destination driven. To put NASA on a better footing to fulfill this vision, the Agency is taking on a major transformation aimed at eliminating stovepipes, becoming more integrated and results oriented, and reducing risks while working more economically, efficiently and effectively. Although NASA is in the very early stages of this transformation, it is already undertaking initiatives to reshape and strengthen its workforce, including developing a strategic human capital plan and an agency-wide workforce planning and analysis system. This subcommittee is also considering legislation proposed by NASA that would provide the Agency with an assortment of tools and authorities to facilitate its efforts to recruit and retain skilled personnel and reshape its workforce.

The Subcommittee asked that we discuss the four major management challenges we identified at NASA in our latest *Performance and Accountability Series* report. These include: (1) strengthening human capital; (2) controlling International Space Station costs; (3) implementing a faster, better, cheaper approach to space exploration; and (4) correcting weaknesses in contract management. First, however, it is very important to recognize that NASA's efforts to address these challenges and undertake a transformation represent a subset of a larger need to fundamentally transform the Federal Government in light of recent trends and long-range fiscal challenges. In this context, I will discuss some of the essential actions that need to be taken by NASA in order to assure that this transformation will become a reality.

Successfully addressing each of the four challenges will be critical for NASA in making sure that it is equipped to achieve its vision for the future. The first challenge—strengthening human capital—will require a concerted and sustained effort by NASA's leadership to commit to change; develop a strategy that ensures the organization has the appropriate mix of employees to meet future business needs; implement effective approaches for acquiring, developing, and retaining talent; developing and retaining talent; and create a results-oriented culture. The remaining challenges facing NASA—controlling International Space Station costs, implementing a faster, better, cheaper approach to space exploration, and correcting weaknesses in contract management—are equally important to address. Without better oversight and management over its most important programs and acquisitions, NASA's transformation stands to lose credibility and support among its partners in industry, the international community, and academia as well as the support of the Congress.

BACKGROUND

NASA's mission encompasses human exploration and development of space, the advancement and communication of scientific knowledge, and research and development of aeronautics and space technologies. Its activities span a broad range of complex and technical endeavors—from investigating the composition, evaluation, and resources of Mars; to working with its international partners to complete and operate the International Space Station; to providing satellite and aircraft observations of Earth for scientific and weather forecasting purposes; to developing new technologies designed to improve air flight safety. In January 2001, we reported that, overall, NASA spends more than $12 billion annually for goods and services supporting these and other activities, mostly on contracts with businesses and other organizations.[1]

Since 1990, we have periodically reported on government operations that we identified as "high risk," because of their vulnerabilities to fraud, waste, abuse, and mismanagement. Since 1999, we have provided each new Congress with a series of reports entitled *Performance and Accountability Series: Major Management Challenges and Program Risks,* providing a perspective on performance and management challenges across the Federal Government, and updated those operations and programs that we have identified as "high risk."

Our reports have identified a number of major management challenges at NASA. In our last report,[2] issued in January 2001, we identified four challenges that warrant increased NASA attention, including one area—contract management—that we continue to categorize as high risk. These four challenges are still applicable today. We plan to issue our next performance and accountability report in January 2003.

ESSENTIAL ELEMENTS OF MAKING A TRANSFORMATION

NASA's recognition that it needs to make a transformation to a more integrated and results-oriented organization comes amid a period of profound transition for our government. This transition is being driven by a number of key trends, including: global interdependence; diverse, diffuse, and asymmetrical security threats; rapidly evolving science and technology; dramatic shifts in the Agency and composition of the population; important quality of life issues; the changing nature of our economy; and evolving government structures and concepts. These trends present a range of challenges that have no boundaries. These trends also contribute to the huge, long-range fiscal and budget challenge facing the United States.

Given these trends and long-range fiscal challenge, the Federal Government needs to engage in a comprehensive review, reassessment, and reprioritization of what the government does, how it does business, and who does the government's business. For their part, agencies like NASA must re-examine their policies, programs, and

---

[1] U.S. General Accounting Office, *Major Management Challenges and Program Risks: National Aeronautics and Space Administration,* GAO–01–258 (Washington, D.C.: Jan. 2001).

[2] GAO–01–258.

operations as well as their human capital policies and practices. The status quo is simply unacceptable. The long-range numbers do not add up. This re-examination will in turn require federal agencies to transform their cultures and shift their overall orientation from:

- Process to results
- Stovepipes to matrixes
- Hierarchical to flatter and more horizontal structures
- An inward focus to an external (citizen, customer, and stakeholder) focus
- Management control to employee empowerment Reactive behavior to proactive approaches Avoiding new technologies to embracing and leveraging them
- Hoarding knowledge to sharing knowledge
- Avoiding risk to managing risk
- Protecting turf to forming partnerships.

The nature and scope of the cultural transformation that needs to take place in many agencies across the Federal Government will take years to accomplish—easily outrunning the tenures of most political appointees. At the same time, our work over the years has amply documented that many agencies suffer from a range of long-standing management challenges and a lack of attention to basic stewardship responsibilities, requiring concerted action and sustained top-level attention if they are to be addressed.

One option for addressing the issues agencies face is to create a Chief Operating Officer (COO) position for selected departments and agencies that would provide sustained management attention essential for addressing key stewardship responsibilities in an integrated manner while helping to facilitate the transformation process within an agency. The long-term responsibilities, professional and nonpartisan in nature, could include strategic planning, organizational alignment, core values stewardship, human capital strategy, performance management, communications and information technology management, financial management, acquisition management, risk management, knowledge management, matrix management, and change management. Ideally, the COO position should be at the Deputy level, have a term appointment of five to seven years, and be subject to a performance contract.

I testified before the National Commission on Public Service, chaired by Paul Volcker, earlier this week. During my testimony, I noted that agencies that are experiencing particularly significant challenges in integrating disparate organizational cultures along with agencies engaged in major transformation efforts, like NASA, may be especially appropriate first phase candidates for a COO position.[3]

NASA is just beginning to undertake major efforts to transform itself. Mr. O'Keefe came on board as the Agency's new Administrator in January 2002 and began articulating his vision for NASA's future this April. Concurrently, NASA is undertaking human capital and other management initiatives designed to foster better financial management, information technology management, and budget and performance integration. In view of these efforts, the Office of Management and Budget this week identified NASA as leading the government in progressing toward the implementation of five government-wide change initiatives contained in the President's Management Agenda. These include: strategic human capital management; competitive sourcing; improved financial performance; expanded electronic government; and budget and performance integration. While NASA's progress is noteworthy, it is currently rated as a "red light" in each of the five key areas in the President's Management Agenda.

---

[3] U.S. General Accounting Office, *Managing for Results? Using Strategic Human Capital Management to Drive Transformational Change,* GAO–02–940T (Washington, D.C.: July 15, 2002).

STRENGTHENING HUMAN CAPITAL

The Challenge

Like many agencies, NASA is facing substantial challenges in attracting and retaining a highly skilled workforce. Left unchecked, reductions in the space shuttle workforce would have jeopardized NASA's ability to safely support the shuttle's planned flight rate. NASA is taking comprehensive steps to address this problem, but the agency needs to make sure that it can sustain its commitment to implementing a strategic approach to marshaling, managing, and maintaining human capital.

Attracting and retaining a workforce with science, engineering, and technology skills is critical to keeping the space shuttle safe and operational.

Leading public organizations here in the United States and abroad have found that strategic human capital management must be at the centerpiece of any serious change management initiative and efforts to transform the cultures of government agencies. People are an agency's most important organizational asset. They define its culture, drive its performance, and embody its knowledge base.

In January 2001, we designated strategic human capital management as a government-wide high-risk area. As our January 2001 High-Risk Series and Performance and Accountability Series reports make clear, serious human capital shortfalls are eroding the ability of many agencies, and threatening the ability of others, to economically, efficiently, and effectively perform their missions. Plainly, the problem is not federal employees. Rather, the problem is the lack of a consistent strategic approach to marshaling, managing, and maintaining the human capital needed to maximize our government performance and ensure its accountability. Our High-Risk report outlined four pervasive human capital challenges now facing the Federal Government.

- Leadership, continuity, and succession planning
- Strategic human capital planning and organizational alignment
- Acquiring and developing staffs whose size, skills, and deployment meet agency needs
- Creating results-oriented organizational cultures.

As we reported in January 2001, the shuttle workforce had declined significantly in recent years to the point of reducing NASA's ability to safely support the program. Many key areas were not sufficiently staffed by qualified workers, and the remaining workforce showed signs of overwork and fatigue. To the Agency's credit, NASA has recognized the need to revitalize the shuttle's workforce, discontinued its downsizing plans for the shuttle program in December 1999 and initiated efforts to hire new staff. In September 2001, we testified that NASA was hiring approximately 200 full-time equivalent staff and it had focused more attention on human capital in its annual performance plan by outlining an overall strategy to attract and retain a skilled workforce. But even with these gains, there were still considerable challenges. For example, NASA's new staff would require considerable training and the Agency still needed to deal with critical losses due to retirements in coming years.

NASA believes that similar workforce problems affect the entire agency. The average age of its workforce is over 45, and the Agency is finding it particularly difficult to hire people with engineering, science, and information technology skills—fields critical to NASA missions. At this time, within the science and engineering workforce, the over-60 population outnumbers the under-30 population nearly three to one. Currently, 15 percent of NASA's science and engineering employees are eligible to retire; within five years, about 25 percent will be retirement eligible. At the same time, the pipeline of people with science and engineering skills is shrinking. According to NASA's Inspector General, the Agency is also facing the loss of significant procurement expertise through the year 2007.

NASA is taking steps to address its workforce predicament. For example, it is developing an agency-wide integrated workforce planning and analysis system as part of its new financial management system. This system is expected to track the distribution of NASA's workforce across programs, capture critical competencies and skills, determine management and leadership depth, and facilitate gap analyses.

NASA has also developed a strategic human capital plan, which incorporates strategies, tactical actions, and metrics to support human capital goals. The plan has been submitted to the Office of Management and Budget and the Office of Personnel Management for review. The plan is based on a planning model developed by the Office of Personnel Management (OPM) as well as our own model, which we published in March 2002.[4]

Our model is designed to help agency officials effectively lead and manage their people and integrate human capital considerations into their daily decision-making and the program results they seek to achieve. In doing so, the model highlights the importance of a sustained commitment by agency leaders to maximize the value of their agency's human capital and to manage related risks. Consistent with OPM's and the Office of Management and Budget's views, our model of strategic human capital management embodies an approach that is fact-based, focused on achieving strategic results, and incorporates merit principles and other national goals.

Additionally, NASA has renewed attention to hiring applicants just out of college and intends to pursue this even more aggressively in coming years. It is also undertaking a number of initiatives and activities aimed at acquiring and retaining critical needed skills, such as using the new Federal Career Intern Program to hire recent science and engineering graduates, supplementing the workforce with nonpermanent civil servants where it makes sense, and implementing a program to repay student loans to attract and retain employees in critical positions.

This subcommittee is currently considering a series of legislative proposals developed by NASA to provide it with further flexibilities and authorities for attracting and retaining a skilled workforce. These include streamlining hiring procedures; making noncompetitive conversions of term employees to permanent positions; offering larger recruitment, relocation, and retention bonuses; expanding use of early retirement; and providing authority for permanent and enhanced buyouts.

Several of the NASA issues mirror aspects of other legislative proposals such as the Federal Human Capital Act.[5] While we have not performed a detailed analysis of the support behind NASA's legislative proposals, several points are worthy of consideration.

- First, before agencies embark on major changes to their human capital management strategy, they must come to grips with developing a realistic picture of how they can reconcile their wants, needs, and affordabilities. This will require difficult tradeoffs.
- Second, the addition of flexibilities and authorities alone will not solve an agency's workforce problems. Agencies need to undertake a wide array of initiatives to attract, retain, and motivate a top quality workforce. These include such actions as revitalizing recruiting and college relations efforts; conducting employee feedback surveys to set priorities and assess progress; conducting employee preference surveys so employees can be given the opportunity to work in areas that interest and energize them consistent with overall institutional needs; inventorying the skills and knowledge of existing employees; initiating professional development programs for newly hired staff to help them transition and progress; implementing modern, effective, and credible performance appraisal and management systems; redesigning training programs to directly link them to core competencies; and implementing employee-friendly benefits, such as day care centers, business casual dress, flex time, and public transportation subsidies.
- Third, agencies need to make the most of current flexibilities and authorities already available. These flexibilities are identified by OPM in its guide, *Human Resource Flexibilities and Authorities in the Federal Government.*

[4] U.S. General Accounting Office, *A Model of Strategic Human Capital Management,* GAO–02–373SP (Washington, D.C.: March 2002).

[5] We previously testified on this proposed legislation in March 2002. See U.S. General Accounting Office, *Managing for Results: Building on the Momentum for Strategic Human Capital Reform. GAO–02–528T (Washington, D.C.: March 2002).*

They include such things as the ability to use commercial recruiting firms to recruit for vacancies; customize merit promotion plans and performance systems; increase basic pay to attract and retain staff with unusually high or unique qualifications; and grant substantial cash incentive awards. Agencies should develop a sound business case for using these flexibilities by focusing on how a given flexibility will address human capital challenges and ultimately improve agency results. In tandem with exercising these flexibilities, agencies must learn to effectively balance its pay and incentive programs to encourage both individual and team contributions to achieving results.

- Fourth, agencies need effective succession planning. NASA's workforce profile, particularly for science and engineering workers, points to the need for this. Faced with the same problems at GAO, we reinstated our Executive Candidate Development Program, under which candidates are selected through a rigorous competitive process and are prepared for assignments at the SES level. While the potential loss of expertise through retirements will be substantial, this turnover also affords NASA's Administrator the opportunity to change culture, skill mix, deployment locations, and other agency attributes. NASA will, however, need to leverage technology and enhance its training efforts to help make this transition and facilitate needed knowledge sharing initiatives.
- Fifth, agencies must ensure that strategic human capital plans are results-oriented and data driven. This includes developing appropriate information on the number and location of employees and their key competencies and skills as well as data on the profile of the workforce, performance goals and measures for human capital approaches. Further, agencies must effectively use this data to develop strategies that continually ensure they have the right mix of employees to meet its future needs. A key to success in this area will be NASA's ability to implement its new financial management system, since it will encompass the new workforce planning and analysis system.

  Instituting a results-oriented culture involves fostering a collaborative environment where managers, teams, and employees are empowered to accomplish programmatic goals. It also includes creating a performance management system that provides candid and constructive information to individual employees, objective and fact-based information to managers, and the information and documentation necessary to deal with poor performers.

  Modernizing agency performance appraisal and management systems and linking them to strategic plans and desired outcomes should be a top priority. Leading organizations use their performance management systems as a key tool for aligning institutions, unit, and employee performance; achieving results; accelerating change; managing the organization on a day-to-day-basis; and facilitating communication through the year so that discussions about individual and organizational performance are integrated and ongoing. To be successful in doing this, the performance management system must link pay and incentive programs to individual knowledge, skills, and abilities and contributions to achieving organizational results.
- Lastly, it is critical for agencies to sustain commitment to embracing human capital management. Agency leaders need to see people as vital assets to organizational success and must invest in this valuable asset. Agencies can foster this thinking and, commitment in their future leaders through efforts such as succession planning and executive development. In addition, agencies need to hold managers accountable for effectively managing people and actively supporting these concepts. In NASA's case, the importance of the Administrator's personal commitment to change the workforce as well as the way the Agency does business cannot be overstated. His leadership and commitment is essential, but he will need help to be successful, particularly from managers at NASA centers in order to overcome resistance to share knowledge and new ideas.

CONTROLLING SPACE STATION COSTS

The Challenge
Development costs for the International Space Station have soared to the point where NASA has had to make substantial cutbacks in the program. This has negatively impacted NASA's credibility with Congress and raised concern among international partners and the scientific community about the viability of the space station. NASA is taking action to keep costs in check, but its success in this area still faces considerable challenges.

Cost overruns have hampered efforts to build and operate the International Space Station

The International Space Station is characterized as one of the most challenging engineering feats ever attempted. It also represents an important effort to foster international cooperation in scientific research and space exploration. But NASA has been facing considerable difficulties in controlling costs and maintaining the scheduling. The cost to complete assembly has mushroomed by about $5 billion to the current estimate of about $30 billion, and while assembly of the station was originally expected to be completed in 2002, NASA now expects it to be done in 2006. As a result, NASA has had to make substantial cutbacks in the program, which in turn, has raised concern among NASA's international partners and the scientific community about the viability of the space station. The future of the space station program hinges on NASA's ability to stem cost growth and schedule delays and to reestablish its credibility with Congress, NASA's international partners, and the scientific community.

NASA has had difficulty predicting and controlling costs and scheduling for the space station since its inception in 1984. In September 1997, we reported that the cost and schedule performance of the space station's prime contract, which showed signs of deterioration in 1996, had continued to worsen steadily and that program financial reserves for contingencies had deteriorated, principally because of program uncertainties and cost overruns. In our January 2001 Performance and Accountability Series report, we reported that the prime contract for the space station was initially expected to cost over $5.2 billion, and the assembly of the station was expected to be completed in June 2002. But by October 2000, the prime contractor's cost had grown to about $9 billion, of which $986 million was for cost overruns, and the station was not expected to be complete until April 2006. NASA's Office of Inspector General (OIG) reported the same cost overrun in a February 2000 audit report, and based on recommendations in that report, NASA agreed to take several actions, including discussing the prime contractor's cost performance at regularly scheduled meetings and preparing monthly reports to senior management on the overrun status.

Our recent work shows that the reasons for continued cost growth include inadequate definition of requirements, changes in program content, and schedule delays and inadequate program oversight.[6] NASA has controls in place that should have alerted management to the growing cost problem and the need for mitigation, but these were largely ignored because of NASA's focus on fiscal year budget management rather than on total program cost management.

The estimated cost growth is having a profound effect on the utility of the space station—with substantial cutbacks in construction, the number of crew members, and scientific research. As a part of the space station restructuring, further work and funding for the habitation module and crew return vehicle have been deferred, thus requiring the on-orbit crew to be reduced from seven to three members. This will limit the crew member hours that can be devoted to research. Additionally, NASA has cut back on the number of facilities available for research—from 27 to 20. This will eliminate some experiments, such as those relating to biotechnology.

[6] U.S. General Accounting Office, *Space Station: Actions Underway to Manage Cost, but Significant Challenges Remain,* GAO–02–735 (Washington, D.C.: July 18, 2002).

NASA's international partners and the scientific community are not satisfied with these and other reductions in capabilities and have raised concerns about the viability of the space station science program.

NASA is instituting a number of management and cost estimating reforms. But there are significant challenges to their successful implementation. First, NASA is now preparing a life cycle cost estimate for the program. Completing this may be difficult because NASA's financial management system has proven inadequate for tracking space station costs. Second, NASA is undertaking several studies to see how research can be maximized, but these will not be completed until September 2002, leaving NASA with a small window of opportunity to incorporate their results into the 2004 budget. Third, NASA has not yet reached an agreement with its international partners on an acceptable on-orbit configuration, sharing of research facilities, and the sharing of cost. Thus, the capacity and capabilities of the space station, the scope research that can be accomplished, and the partners' share of operating costs are unknown at this time.

IMPLEMENTING A FASTER, BETTER, CHEAPER APPROACH TO SPACE EXPLORATION

The Challenge

NASA has been following a faster-better-cheaper management philosophy to reduce costs, become more efficient, and increase scientific results by conducting more and smaller missions in less time. While NASA has had many successes, failures of two Mars probes shows that there are limits to this approach, particularly in terms of NASA's ability to learn from past mistakes. NASA has taken steps in recent years to strengthen lessons learning within the agency, but more needs to be done to overcome cultural and organizational impediments.

The Mars Sojourner Rover was built under the faster-better-cheaper philosophy.

Since 1992, NASA has been following a faster-better-cheaper management philosophy to reduce costs, become more efficient, and increase scientific results by conducting more and smaller missions in less time. The faster-better-cheaper approach works by focusing on building less expensive space probes much quicker than in the past. It is intended to stimulate innovative development and application of technology, streamline policies and practices, and energize and challenge a workforce to continue to safely and successfully undertake bold new missions in an era of diminishing resources.

While the approach has been successfully used for numerous missions, the failures of two missions to Mars brought increased scrutiny. The Mars Climate Orbiter, which was intended to observe Mars' seasonal climate and daily weather from a low orbit around the planet, was lost on September 23, 1999. Then on December 3, 1999, the Mars Polar Lander, a robotic spacecraft intended to land near the South Pole of Mars for a planned 90-day mission to study the planet's layered polar terrain, was also lost.

NASA-sponsored investigative boards found that opportunities to identify and resolve problems were missed due to poor communications, budget pressures, and poor management and engineering practices. Upper management officials were not aware of the extent of the programs' problems.

In our January 2001 Performance and Accountability report, we reported that NASA still faced significant challenges to creating highly reliable missions and fostering open communications under the budget constraints of the Agency's faster-better-cheaper space exploration strategy. In addition, success required an integration of lessons learned from failures on an agency-wide basis.

NASA now recognizes the importance of learning from the past to ensure future mission success and uses several mechanisms to capture and disseminate lessons learned. In January 2002, for example, we reported[7] that NASA had developed a Web-based lessons learned database, and used training, program reviews and periodic revisions to agency policies and guidelines to communicate lessons.

However, we also found that these tools were limited in their effectiveness. NASA program and project managers reported to us that they were unfamiliar with lessons generated by other centers and programs and many stated that they were dissatisfied with NASA's lessons learned processes and systems. For example, they were not using NASA's lessons learned information system partly because it was time-consuming to do so.

We also identified problems hampering knowledge sharing within NASA that certainly reflect a need for a significant transformation within the Agency. In particular, many program and project managers told us that they believed senior management support was lacking for sharing lessons learned. There were also significant cultural barriers to knowledge sharing—beyond the difficulties associated with a stovepiped environment. For example, there were no agency-wide incentives for sharing knowledge; many managers simply lacked time to take part in knowledge-sharing activities; and the sharing of lessons learned was not highly valued across the board. Clearly, with the difficulties the Agency is facing in hiring highly skilled employees, leveraging the institutional knowledge of its experienced workforce is critical.

We made several recommendations to address these underlying problems as well as recommendations to improve NASA's current knowledge sharing mechanisms. NASA generally agreed with our recommendations and plans to implement them.

CORRECTING WEAKNESSES IN CONTRACT MANAGEMENT

The Challenge

Much of NASA's success depends on the work of its contractors—on which it spends the greatest part of its funds. But for many years, NASA has not been able to effectively oversee contracts, principally because it has lacked accurate and reliable information on contract spending and it has placed little emphasis on end results, product performance, and cost control. NASA has addressed many acquisition-related weaknesses, but key tasks remain, including implementation of a new integrated financial management system.

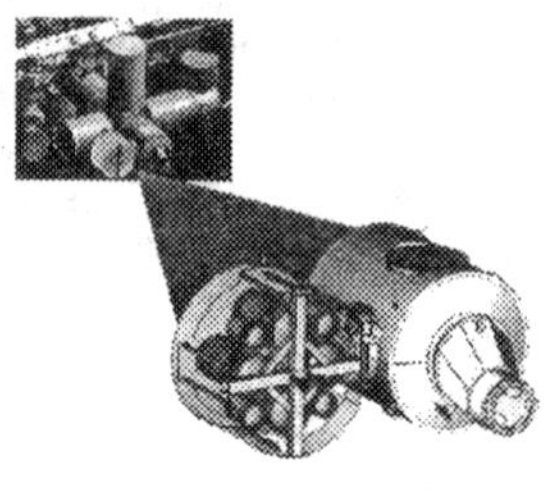

The initial effort to develop the Propulsion Module for the International Space Station was unsuccessful because NASA proceeded with the contractor's proposal without following fundamental contract management processes.

Since 1990, we have identified NASA's contract management function as an area at high risk due to its ineffective systems and processes for overseeing contractor activities. Our reports and testimonies since then, have demonstrated just how debilitating contract management and oversight weaknesses can be to important space programs. Our July 2002 report on the International Space Station, for example, found that NASA did not effectively control costs as well as technical and scheduling risks, provide adequate oversight review, or effectively coordinate efforts with its partners.[8] In other examples, we found that NASA lacked effective systems and processes for overseeing contractor activities and did not emphasize controlling costs. NASA's accounting systems were designed prior to implementation of current federal cost accounting and financial systems that require agencies to track and maintain data for estimating and controlling costs, performance measurement, and making economic trade-off decisions.

In recent years, NASA has made progress in addressing its contract management challenges. In July 1998, for example, we reported that NASA was developing sys-

[7] U.S. General Accounting Office, *NASA: Better Mechanisms Needed for Sharing Lessons Learned*, GAO–02–195 (Washington, D.C.: Jan. 30, 2002).

[8] GAO–02–735.

tems to provide oversight and information needed to improve contract management and that it had made progress evaluating its field centers' procurement activities on the basis of international quality standards and its own procurement surveys. In January 1999, we found that NASA was implementing its new system for measuring procurement-related activities and had made progress in evaluating procurement functions in its field centers.

But much work remains to be done to strengthen contract oversight. A key task is modernizing NASA's financial management systems. According to NASA, the Agency's financial management environment is comprised of decentralized, nonintegrated systems with policies, procedures, and practices that are unique to its field centers. For the most part, data formats are not standardized, automated systems are not interfaced, and on-line financial information is not readily available to program managers. Thus, it is difficult to ensure contracts are being efficiently and effectively implemented and budgets are executed as planned. In addition, NASA has pointed out that the cost to maintain these systems has been high, since both data and software are replicated at each field center.

The inadequacy of NASA's financial management system has further impact. Without a more effective financial management system, NASA will likely continue to have difficulty providing relevant, reliable, timely financial data—including cost information—that can be used on a real-time basis by program managers to monitor costs, schedule, and performance. In March 2002, we testified[9] that NASA was unable to provide us with detailed support for amounts obligated against cost limits established by the fiscal year 2000 NASA Authorization Act. This was due, in large part, to NASA's lack of a modern, integrated financial management system.

To its credit, NASA is working toward implementing an integrated financial management system that it expects to be fully operational in fiscal year 2008 at an estimated cost of $691 million. This is NASA's third attempt toward implementing a new integrated financial management system. The first two efforts were abandoned after 12 years and after spending a reported $180 million. NASA's current approach focuses on learning from other organizations' successes in implementing similar projects, as opposed to revisiting its own failures. NASA has also abandoned the single product approach that the two prior attempts had as their basic architecture. Instead, the project will be broken down into implementable modules on the basis of the availability of proven software products.

Given the high stakes involved, it is critical that NASA's leadership provide the necessary direction, oversight, and sustained attention to ensure that this project is successful. In this regard, NASA's Administrator comes to the position with a strong management background and expertise in financial management. Based on our discussions with the Administrator, he has made clear that he plans to make financial management a top priority.

The task ahead, however, is daunting. In a recent internal review, NASA found that the total cost estimate for deployment of the core financial module at all NASA centers had grown considerably beyond the cost initially contemplated. The review also revealed interoperability and security vulnerabilities within the current information infrastructure. To address these continuing problems, the Administrator appointed an executive to provide leadership and accountability in the direction and operation of the new system. He also recently decided that the near-term focus of the program should be to ensure a successful and rapid deployment of the core financial module—the backbone of the system—and that the schedule of the remaining modules should undergo further risks assessments before moving forward.

While modernizing NASA's financial management system is central to producing accurate and reliable financial information needed to support contract management activities, technology alone will not solve NASA's contract management problems. NASA must also ensure that the cost data collected and maintained in its financial management system are sufficiently detailed to allow comparisons of actual costs to estimates and thereby provide an early warning of cost overruns or other related difficulties. As we reported in August 2001, NASA's management practices and business processes do not always facilitate the development of this type of data. For example, we reported that NASA does not track the actual costs of completed space station components even though it often estimates the cost of these components for planning and budgeting purposes. Also, in programs such as the space station, NASA needs to effectively implement new controls planned to strengthen technical and scheduling reviews as well as risk analyses.

[9] U.S. General Accounting Office, *National Aeronautics and Space Administration: Leadership and Systems Needed to Effect Financial Management Improvements,* GAO–02–551T (Washington, D.C.: March 20, 2002).

We are continuing to monitor NASA's progress in addressing contract management weaknesses. In response to a May 24, 2002 bi-cameral, bipartisan request from the Senate Commerce, Science, and Transportation Committee and the House Science Committee, we are currently assessing the extent to which NASA's management of the financial management system acquisition is in accordance with effective system acquisition practices and is designed to support NASA's decision-making needs and external reporting requirements.

In closing, NASA has a long and proud history, and it does many things well. But, times have changed, and NASA must change with the times in considering what it does and how it does business. Moreover, the Agency is facing management challenges, which if not effectively addressed, stand to hurt NASA's credibility with the Congress and its partners and hamper important space missions. I would like to commend Mr. O'Keefe for recognizing the need to transform and making a personal commitment to the transformation effort. The steps he is taking should lay a sound foundation for change. This is reflected in OMB's recent characterization of NASA as leading the government in its progress implementing the five government-wide initiatives identified in the President's Management Agenda. Clearly, NASA is off to a strong start on what will be a long-term effort. The challenge ahead for NASA will be to maintain the momentum to transform, to effectively use existing and new authorities to strategically manage its people, and to quickly implement the tools needed to strengthen management and oversight.

CONTACTS AND ACKNOWLEDGEMENTS

For further information regarding this testimony, please contact Allen Li at (202) 512–4841. Individuals making key contributions to this testimony include, Jerry Herley, Cristina Chaplain, Shirley Johnson, Charles Malphurs, and Sarah Marquis.

(120170)

The United States General Accounting Office

Comptroller General of the United States

DAVID M. WALKER

David M. Walker became the seventh Comptroller General of the United States and began his 15-year term when he took his oath of office on November 9, 1998. As Comptroller General, Mr. Walker is the nation's chief accountability officer and head of the U.S. General Accounting Office (GAO), a legislative branch agency founded in 1921. GAO's mission is to help improve the performance and assure the accountability of the federal government for the benefit of the American people. Over the years, GAO has earned a reputation for professional objective, fact-based, and nonpartisan reviews of government issues and operations.

The long tenure of the Comptroller General gives GAO a continuity of leadership and independence that is rare within government. Both elements help to allow GAO to consider long-range and cross-governmental issues and alert policymakers to problems looming on the horizon, such as the growing burden of entitlement programs or the nation's deteriorating infrastructure.

Before his appointment as Comptroller General, Mr. Walker had extensive executive level experience in both government and private industry. Between 1989 and 1998, Mr. Walker worked at Arthur Andersen LLP, where he was a partner and global managing director of the human capital services practice based in Atlanta, Georgia. He was also a member of the board of Arthur Andersen Financial Advisors, a registered investment advisor. While a partner at Arthur Andersen, Mr. Walker served as a Public Trustee for Social Security and Medicare from 1990 to 1995. Before joining Arthur Andersen, Mr. Walker was Assistant Secretary of Labor for Pension and Welfare Benefit Programs from 1987 to 1989 and in 1985, was Acting Executive Director of the Pension

Benefit Guaranty Corporation. His earlier technical, professional, and business experience was with Price Waterhouse, Coopers & Lybrand and Source Services Corporation, an international human resources consulting and search firm.

Mr. Walker serves as Chair of the U.S. Intergovernmental Audit Forum, the U.S. Joint Financial Management Improvement Program, and the Center for Continuous Auditing. He is on the Board of the International Organization of Supreme Audit Institutions and various educational and not-for-profit entities. He is a Fellow of the National Academy of Public Administration and an active member of various professional public service, and other organizations. Mr. Walker is also listed in Who's Who in the World and Who's Who in America.

Mr. Walker is the author of Retirement Security: Understanding and Planning Your Financial Future (John Wiley & Sons, 1996) and a co-author of Delivering on the Promise: How to Attract, Manage and Retain Human Capital (Free Press, 1998). He has also written numerous articles and opinion letters on a variety of subjects. Mr. Walker is frequently quoted on a range of government and management issues and has been the subject of several cover stores in various national, professional and governmental journals.

Mr. Walker is a certified public accountant. He has a B.S. degree in accounting from Jacksonville University, a Senior Management in Government Certificate in public policy from the John F. Kennedy School of Government at Harvard University, and a Honorary Doctorate in Business Administration from Bryant College. He is married to the former Mary Etheredge, and they have two adult children—a daughter, Carol, and a son, Andy.

Accountability ★ Integrity ★ Reliability

Chairman ROHRABACHER. Thank you. I think the Long March is a Chinese rocket, however. Our next witness is NASA Administrator Sean O'Keefe. I might add that our last NASA Administrator really wanted to be Irish, and those of us who knew him and liked the last Administrator really had fun with that. So just as our President is the first Chief Executive with an MBA, Sean is our first NASA Administrator with a degree in public administration. That education is being put to the test, and so we are very appreciative of the challenges that you have accepted to take on and anxious to hear about your progress today. Thank you, Mr. O'Keefe, you may proceed.

STATEMENT OF HON. SEAN O'KEEFE, ADMINISTRATOR, NATIONAL AERONAUTICS AND SPACE ADMINISTRATION

Mr. O'KEEFE. Well, thank you, Mr. Chairman, and thank you Members of the Committee for calling this hearing, and your expeditious consideration of the proposal that we have advanced just a matter of weeks ago. We are very grateful to you for giving the consideration of the efforts we are underway with. First and foremost, if you permit me, Mr. Chairman, I want to recognize that as illustrative of the observations that Congresswoman Jackson Lee observed, that part of our efforts in dealing with that next generation of explorers are here with us today. The Ames Astrobiology Academy has a dozen folks here from the Silicon Valley area that are visiting Washington during this time. I want to thank them for attending and having the interest in the public policy deliberations here to have an interest in doing so.

They are undergraduates pursuing science, math and engineering-related degrees, and so as a consequence they are the group we hope to influence for future decisions as well. As well, as I want to thank my good friend, Dave Walker, for documenting I think over time, as well as here this morning, the human capital challenges we are facing across the Federal Government of which NASA is not unique in this regard at all, but indeed I think illustrative of some of the issues we will be confronting.

In that regard, I think GAO's decisions and the Comptroller General's decision to make human capital a high risk list item is indicative of the fact that across the federal establishment there is a range of challenges we need to confront and certainly we need to go about them as one agency at a time and with some federal-wide objectives that guide them.

NASA DEPUTY ADMINISTRATOR AS CHIEF OPERATING OFFICER

In that regard as well too, I want to advise the Comptroller General today, this afternoon, we have a nominee before the Senate Commerce Committee to be considered by the President's nomination to be the Deputy Administrator. Fred Gregory a career astronaut and former Air Force officer who has served in the public service for the last dozen years in a variety of senior positions. He is the quintessential Chief Operating Officer candidate, who will be designated as such, assuming the Senate concurs with the President's nomination, so we urge their expeditious consideration of him as well.

PRESIDENT'S MANAGEMENT AGENDA

The President's management agenda identifies much that has been identified by the Comptroller General. Human Capital is one of the top five issues that need to be confronted government-wide and for which government-wide initiatives are encouraged, as well as individual agency-specific activities need to be focused as well. Our efforts at NASA are to address this challenge with a series of initiatives to confront the trend that we are uniquely experiencing that is indicative of that government-wide challenge.

FACTORS INFLUENCING SCIENCE AND ENGINEERING WORKFORCE

Here are just a couple of representative factors that influence our decision to move in the direction we proposed and urge your consideration of those proposals.

We have, to put it I think as charitable as possible, an extremely mature workforce. We have three times as many folks in the science and engineering community who are over 60 as under 30. Twenty-five percent of the science and engineering community workforce will be eligible to retire within the next three to five years. Concurrently, there is a declining shortage of science and engineering disciplines, yet more than half of our very important public servants who are engaged in the activities of science and engineering have Masters and Doctoral degrees, yet at the same time we are seeing a declining trend that has been well-documented by the Hart Rudman Commission, as well as others, in terms of what that shortage has to draw from that we will seek to recruit from in the years ahead.

So as a consequence, we have a convergence of both demand, I think that has been represented and observed by several of the opening statements here this morning by the Members of the Committee, as well as a decline in the interest level that we see across the country in science, engineering, and technology related fields. Indeed, I was struck by the statistic I saw the other day that there are more folks who graduated last month with sports and physical sciences-related kind of activities, exercise sciences-related degrees than in electrical engineering last year.

NASA'S PLANS TO DEAL WITH NATIONAL TRENDS WITH SCIENTISTS AND ENGINEERS

So as a consequence we are seeing a continuing trend in a direction that bodes very poorly in terms of our capacity to actually recruit and retain competent science and engineering professionals that approach the level of competence that we see today.

So our purpose is fourfold. First, it is to use the existing authorities that we have available to us, and the Comptroller General has been an advocate of this approach for some time. We have adhered to that method. We are trying to maximize our leverage using all the authorities provided not only through Title 5 and the very creative approaches that can be employed using those existing authorities, but also those in the Space Act of 1958, as amended. So we are trying to vigorously utilize the capabilities before us.

The second thrust or approach that we are using as an initiative is the development of an agency-wide human capital strategic plan, and I am delighted that the Comptroller General observed that not only is that completed, but we are on our way to implementing that effort as it has been conducted and worked through with the Administration to begin targeted hiring objectives, looking at professional development strategies, workforce shaping techniques that draw the best benchmarking we have seen across federal public agencies over the last several years, managing full-time equivalents on an agency-wide basis as opposed to setting targets at each center and location as if somehow that had a direct relationship. We are trying to manage that in a more proactive way, and emphasizing important diversity objectives, of which the science and engineering community is woefully deficient in terms of our representation of the general population at large.

The third major initiative that we are after is the NASA mission and vision itself. To understand and protect our home planet, to explore the universe and search for life, but third, and what has particular bearing on this question is to inspire that next generation of explorers through education initiatives. We are stepping up that activity and trying to emphasize that area as a dominant theme and approach that we will be incorporating very specifically and proactively as part of our internal fiscal year 2004 deliberations that are currently ongoing.

And, fourth, we are asking for legislative authority. It has been proposed here by the Administration based on the Managerial Flexibilities Act that was forwarded to the Congress last October. It is on the basis of that. The areas that are emphasized here that we will explore, I am sure, this morning, is to recruit experienced scientists and engineers for mid-level entry, to recruit to achieve diversity objectives, to retain high performers, to use innovative techniques which have been pilot tested in other agencies previously and extend those authorities to NASA specifically, to expand authorities while remaining consistent with principles of merit protection, performance objectives, and establish public policy for selective preference in various areas.

We urge your support for these initiatives and look forward to exploring them. And again, Mr. Chairman, my thanks for your expeditious consideration of these proposals.

[The prepared statement of Mr. O'Keefe follows:]

PREPARED STATEMENT OF SEAN O'KEEFE

I am pleased to appear before the Subcommittee today to discuss NASA's Human Capital challenges. The Agency faces a number of strategic threats to our ability to manage our Human Capital effectively and efficiently. We are asking for a number of legislative provisions that will give our managers the tools they need to reshape and reconstitute the workforce to respond to these threats.

When President Eisenhower and the Congress created NASA, they sought to establish a government agency that could undertake and overcome the Nation's technological challenges in aeronautics and space exploration. Without NASA, there would be no American presence to take up these challenges. During the Cold War, the very best minds of our nation joined forces to transform the futuristic dreams of our parents' generation into the historical reality our children learn about in today's classrooms. The legacy of that work continues today. Across the Nation, NASA scientists, engineers, researchers, and technicians have made and continue to make remarkable discoveries and advancements that touch the lives of every American. We are an Agency committed to "pioneering the future" as only NASA can.

One of the greatest challenges before the Agency today is having the people—the human capital—available to forge ahead and make the future breakthroughs tomorrow's everyday reality. NASA's history is celebrated worldwide for having accomplished the things that no one has ever done before. None of those achievements happened by accident. They were the result of management innovation, revolutionary technologies and solid science and research. These three pillars of NASA's achievement were built by the men and women of NASA and without them, the history of achievement that we celebrate in aeronautics and space exploration never would have been possible. History is made everyday at NASA; but to maintain our leadership position, a new generation must be forged to carry our nation's innovation and exploration forward.

Vision And Mission

When I assumed the leadership of NASA late last year, I wanted to ensure that this pathfinder agency had the means and mission to support that pioneering spirit through the next several decades. NASA has a vital role to play in today's world and my testimony today will touch on the management challenges that NASA must overcome if we are to achieve our mission. NASA is intent on continuing the gains made over 44 years while pushing the edge of the envelope of what appears today to be impossible. We have developed a roadmap to continue our work in a more efficient, collaborative manner. NASA's imperative is not only for the sake of human knowledge—it is for our future and our security.

Soon after I came to NASA, we developed a new strategic framework and vision for the Agency. It is a blueprint for the future of exploration and a roadmap for achievement that we hope will improve the lives of everyone in this country and everyone on this planet. Our new vision is to improve life here, to extend life to there, and to find life beyond. This vision frames all that we do and how we do it. NASA will do this by implementing our mission—to understand and protect our home planet; to explore the Universe and search for life; to inspire the next generation of explorers. . .as only NASA can.

To understand and protect our home planet, NASA will work to develop and employ the technologies that will make our Nation and society a better place. We will work to forecast the impact of storms on one continent upon the crop production on another; we will work to trace and predict the patterns of mosquito-borne diseases, and study climate, geography and the environment—all in an effort to understand the multiple systems of our planet and our impact upon it.

Our mission's second theme is to explore the universe and search for life. NASA will seek to develop the advanced technologies, robotics, and science that eventually will enable us to explore and seek firsthand the answers and the science behind our most fundamental inquiries. If we are to achieve such ambitious objectives, there is much we still must learn and many technical challenges that must be conquered.

For example, today's rockets that have been the engine of exploration since the inception of space travel are today at the limit of what they can deliver. Our current budget before the Congress invests nearly $1 billion over five years for a nuclear systems initiative as a first step in addressing this challenge. Propulsion is only one of the challenges facing further human exploration of space. The physical challenges incurred by our space explorers also must be better defined. We still do not know or understand the long-term effects of radiation and exposure to a microgravity environment upon the human body. The infant steps we have taken via the Space Shuttle and the International Space Station have given us many answers to explore, but they have yielded even more questions for us to consider.

Our third mission objective is to inspire the next generation of explorers. America often looks to NASA to help our Nation build an unequalled pool of scientific and technical talent. NASA accepts that responsibility and in partnership with the U.S. Department of Education, other federal agencies, and industry and educational partners, we will work to motivate our nation's youth to embrace the study of mathematics, science and engineering disciplines. Without the scholars to take the study of these disciplines to their next level, the missions we seek to lead remain bound to the launch pad. As the U.S. Department of Labor has reported, the opportunities in the technology sector are expected to quadruple in this decade. Unfortunately, the pool of college students enrolled in mathematics, science and engineering courses continues to decline. NASA faces similar challenges with having the scientific and engineering workforce necessary to fulfill its missions.

Our mission statement concludes with the statement, "as only NASA can." Our agency is one of the Nation's leading research and technology federal agencies. We possess some of our nation's most unique tools, capabilities and expertise and represent a national asset and investment that is unparalleled in the world. Nonetheless, to achieve success in our mission, our activities must focus on those areas where NASA can make unique contributions. To make the best use of our workforce and other resources, we must also leverage the unique contributions of our partners in academia, industry, and other federal agencies.

Our commitment to the American taxpayer is to continue providing a direct and very tangible means of improving life on our planet. Extending life beyond the reaches of our Earth is not a process driven by any particular destination. Rather it is driven by science that will contribute to the social, economic, and intellectual growth of our society and the people who make that science possible are our greatest asset.

Workforce Challenges

NASA's ability to fulfill its ambitious mission is dependent on the quality of its workforce. An agency is only as strong as its people. To be successful, they need to be world-class if they are to be expected to break new ground in science and technology, explore the universe, or pioneer exciting discoveries here on Earth and beyond. NASA needs the best and the brightest. This means that NASA needs, not only the scientists and engineers who form the core of our workforce, but also highly competent professionals who can support NASA's technical programs, and address the Agency's financial, human capital, acquisition, and business management challenges.

Today, NASA faces an increasing management challenge in attracting, hiring, and retaining the talented men and women who, inspired by our amazing discoveries and innovations of the past four decades, will help mold the future of our nation's aeronautics and space programs. As a nation, we must ensure that the Agency continues to have the scientific and technical expertise necessary to preserve our role as the world's leader in aeronautics, space and Earth science, and emerging technology research. The President already has indicated his commitment to the strategic management of human capital in the federal workforce, by making this issue number one in his Management Agenda. In fact, the President's Management Agenda specifically references the human capital challenge that NASA faces and related skill imbalances. The President's recognition of the challenge NASA faces is shared by the General Accounting Office, which has placed the management of human capital as one of the items on the government-wide "high-risk list."

At NASA, we are ready to do our part to make sure that we have the best people for the job at hand, and to do that we need to manage this resource efficiently and responsibly, as well as compete favorably in a very competitive market place. We have developed a Strategic Human Capital Plan to establish a systematic, agency-wide approach to human capital management, aligned with our vision and mission. The Plan assesses NASA's current state with respect to human capital management, then goes on to identify goals, problems, improvement initiatives, and intended outcomes. The plan is an integrated approach to address the concerns of the Administration as well as our internal human capital needs.

NASA's ability to implement its mission in science, technology, and exploration depends on our ability to reconfigure and reconstitute a world-class workforce. The human capital flexibilities that we are requesting, consistent with the Administration's Managerial Flexibility Act, will help us shape the workforce necessary to implement our mission today and in the future. We currently face many skill mix imbalances that impede our ability and it will be even more challenging in the future if we fail to act now.

The Committee's letter of invitation requested that NASA address Shuttle competitive sourcing, workforce adjustments to the Space Station program and the Strategic Resources Review (SRR). There are a number of reviews which are ongoing that are addressing each of these issues. At the present time, no decisions have been made, nor will be until the studies are complete. The human capital legislative provisions NASA is requesting are based on the Agency's need to rebalance the skill mix of our workforce and to enable us to attract and retain a world class workforce for the future. Regardless of the outcome of the studies, we need legislatives measures now to address current workforce skills issues.

Today, NASA's ability to maintain a workforce with the talent it needs to perform cutting-edge work is threatened by several converging trends. Each trend in isolation is concerning; in concert the indicators are alarming. We need to address these trends now by anticipating and mitigating their impact on NASA's workforce in the near-term and beyond. These indicators could lead to a severe workforce crisis if we do not take prompt action. The warning signs are here, and we are paying attention. Many of our planned actions to deal with threats to our human capital are possible without the aid of Congress; but some of the solutions require legislation. We are proposing a number of human capital provisions, which I believe are crucial steps toward averting a workforce crisis.

The trends I'd like to discuss with you today fall into two broad categories. First, there are trends that affect the nationwide labor market, and the applicant pool from which we draw our workers. These indicators affect other employers, not just NASA, and point to worsening employee pipeline issues in the future. Secondly, I would like to address a number of NASA-specific demographics. Coupled with the nationwide issues we face, the NASA picture shows us that we need to take action and take it now.

Nationwide Trends

The Shrinking Scientist and Engineer (S&E) Pipeline

There is growing evidence that the pipeline for tomorrow's scientists and engineers is shrinking. We are facing a critical shortage of students pursuing degrees in disciplines of critical importance to NASA—science, mathematics, and engineering. Several recent National Science Foundation reports document a disturbing trend: the science and engineering (S&E) pipeline has been shrinking over the past decade. This trend begins at the undergraduate level and extends through the ranks of doctoral candidates. Here are some statistics that illustrate what currently is happening to the S&E pipeline:

- Undergraduate Engineering Enrollment—The number of students enrolling in *undergraduate* engineering decreased by more than 20 percent between 1983 and 1999. [National Science Board, Science and Engineering Indicators-2002, Arlington, VA: National Science Foundation, 2002 (NSB–02–01)]

Undergraduate Engineering Enrollment Trend

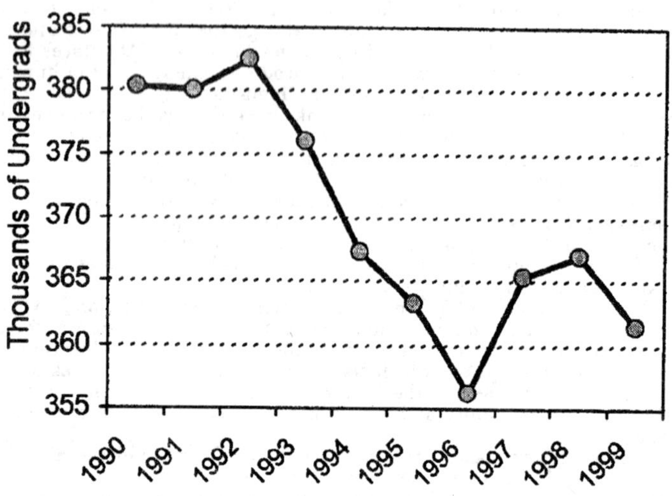

- Graduate S&E Enrollment—Engineering *graduate* enrollment also declined from a high in 1992 of 128,854 to 105,006 in 1999. Graduate enrollment in the physical sciences, earth sciences, and mathematics also showed a downturn between 1993 and 2000. [National Science Foundation Data Brief, Growth Continued in 2000 in Graduate Enrolment in Science and Engineering Fields (NSF–02–306), December 21, 2001]
- Post-Graduate S&E Enrollment—By the year 2000, the number of doctorates awarded annually in engineering had declined by 15 percent from its mid-decade peak; since 1994, the number of doctorates in physics declined by 22 percent. Even in mathematics and computer science—where job opportunities are on the rise—the number of doctorates awarded declined in 1999 and 2000. [National Science Foundation, Info Brief Declines in U.S. Doctorate Awards in Physics and Engineering (NSF–02–316), April 2002]
- Foreign S&E Enrollment—40 percent of the graduate students in America's engineering, mathematics, and computer science programs are foreign nation-

als. In the natural sciences, the number of non-citizens is nearly one in four. When we concentrate on engineering graduate students who are U.S. citizens, the number of enrollees declined precipitously between 1993 and 1999: from more than 77,000 to just over 60,000, a 23 percent drop in under a decade. [National Science Board, Science and Engineering Indicators-2002, Arlington, VA: National Science Foundation, 2002 (NSB–02–01)]

- Aerospace Enrollment—Graduate enrollment in aerospace engineering has declined steadily in recent years—from 4,036 in 1992 to 3,407 in 2000, pointing to a diminishing interest in aerospace as a career. [National Science Board, Science and Engineering Indicators-2002, Arlington, VA: National Science Foundation, 2002 (NSB–02–01) and National Science Foundation Data Brief, Growth Continued in 2000 in Graduate Enrolment in Science and Engineering Fields (NSF–02–306), December 21, 2001)]

NASA is not alone in its search for enthusiastic and qualified employees. Throughout the Federal Government, as well as the private sector, the challenge faced by a lack of scientists and engineers is real and is growing by the day. The situation is summarized in the recently issued Hart-Rudman Commission's Final Report, "The harsh fact is that the U.S. need for the highest quality human capital in science, mathematics, and engineering is not being met."

The nationwide trends I have described have great significance to NASA since the Agency relies on a highly educated science and engineering workforce: nearly 60 percent of the total NASA workforce is S&E, and fully half of those employees have Masters or Doctorate degrees.

Increased Competition for Technical Skills

At the same time that the national S&E pipeline is shrinking, *the demand for the technical skills NASA needs is increasing.* The job market in the S&E occupations is projected to increase dramatically over the next ten years. The need for technical expertise no longer is confined to the technical industries that have been traditional competitors. NASA will face competition from new arenas as graduates in the S&E fields now are sought after by the banking industry, entertainment industry, and elsewhere in career fields not traditionally considered career fields for technical graduates. In the academic sector, traditionally not a competitor, we find ourselves vying for the same high-level technical workers. America's top schools now offer very competitive salaries to academicians with world-class skills—the same skills NASA seeks. Specifically, here are some of the trends that the Nation is seeing in the job market:

- Increasing S&E Positions—The Bureau of Labor Statistics projects that employment in the fields of science and engineering is expected to increase about three times faster than the rate for all occupations between 2000 and 2010, mostly in computer-related occupations. Increases in engineering and the physical sciences are projected at 20 percent and 15 percent, respectively. [National Science Board, Science and Engineering Indicators-2002, Arlington, VA: National Science Foundation, 2002 (NSB–02–01)]

Nationwide Opportunities through 2010

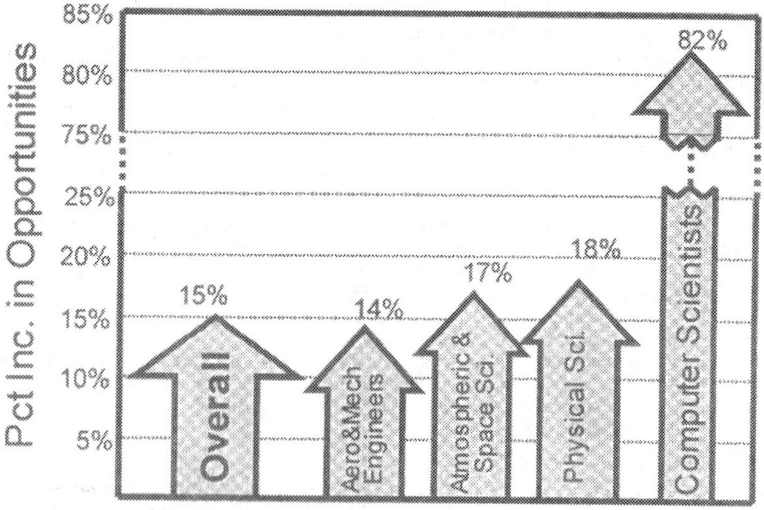

- Increasing S&E Retirements—This report also notes that with current retirement patterns, the total number of retirements among S&E-degreed workers will increase dramatically over the next 20 years. More than half of S&E-degreed workers are age 40 or older, and the 40–44 age group is nearly 4 times as large as the 60–64 age group. As employers seek to fill vacancies created by these retirements, *competition for quality S&E workers will intensify.*
- Low Interest in Government Employment—According to a recent (October 2001) Hart-Teeter poll, the lowest levels of interest in government employment were found among college-educated and professional workers. Only 16 percent of college-educated workers express significant interest in working for the Federal government, and a like number of professionals and managers would opt for a government job. In contrast, the poll also revealed that positive perceptions of private sector work increased dramatically among those with formal education. This indicates that NASA will face a significant challenge in trying to attract experienced mid- and senior-level professionals to the Agency.

Lack of Diversity in the Nationwide Applicant Pool

NASA continues to face challenges in its efforts to recruit scientists and engineers from a candidate pool that is representative of the Nation's diversity. Our missions are intended to serve the interests of humanity and that means building a mission-oriented team that represents the best and brightest of America As the Bureau of Labor Statistics Report, "Working in the 21St Century," has noted, "minorities are the fastest growing part of the labor force." Unfortunately, the S&E labor pool does not follow this trend.

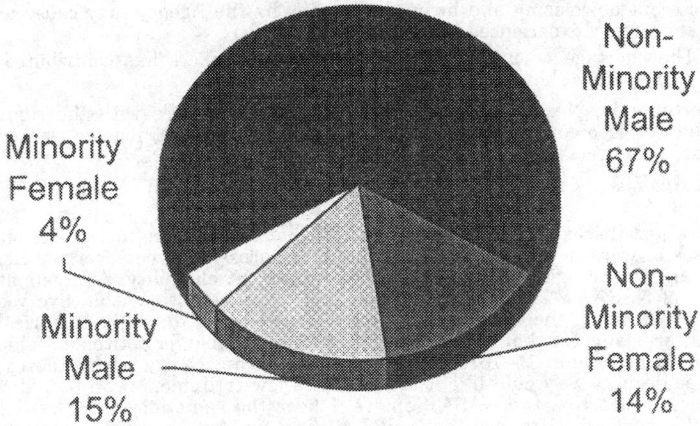

At this time, the diversity of the NASA S&E workforce mirrors that of the U.S. S&E workforce—it is predominantly comprised of white males. Currently, 80% of the S&E workforce within NASA is white; 4.7 percent is Hispanic; 5.7 percent is African-American, and 7.3 percent is Asian/Pacific Islander. Female representation is 18 percent. Since the undergraduate S&E pool also lacks diversity to the same or greater extent, NASA is unlikely to improve upon these demographics without new tools, flexibilities, and strategies. Employers from private industry and academia as well as the public sector all wish to achieve diversity within their workforce, so competition for a relatively small pool of minority and female scientists and engineers is keen. NASA must struggle to compete with these employers, many of whom have the means to offer more competitive job offers to achieve their diversity goals.

NASA Demographics and Trends

Current Skills Imbalances, Gaps, and Lack of Depth Within the NASA Workforce

The trends I have just outlined are not unique to NASA; we share them with other employers in the labor market today. Unfortunately, the difficulties they present to NASA's ability to manage our human capital are only exacerbated by several agency-specific threats, warning us that we need to pay attention to these indicators before they result in a crisis. The challenge of acquiring and retaining the right workforce is not a problem of the future—it exists now. The Agency currently has skill gaps in areas such as nanotechnology, systems engineering, propulsion systems, advanced engineering technology, and information technology. In emerging technology areas, NASA projects the need to employ more civil servants in "hard to fill" areas such as astrobiology, robotics, and fundamental space biology. In other professional areas such as financial management, acquisition, and project management, a lack of depth is becoming detrimental to our ability to manage our resources and programs.

NASA's has undergone significant downsizing over the past decade, reducing its workforce from approximately 25,000 civil servants in FY 1993 to slightly more than 18,000 by FY 2002. NASA made every effort to retain key skills, but in order to avoid involuntary separations in achieving those reductions, it was not always possible to control the nature of the attrition. Inevitably, we lost some individuals with skills we couldn't afford to lose, and now these skills need to be replaced. Through downsizing and the natural attrition process, we lost key areas of our institutional knowledge base.

The 2001 report of the Aerospace Safety Advisory Panel made specific references to NASA's skills deficiencies when they noted the following:

- NASA faces a critical skills challenge in the Shuttle and International Space Station programs despite resumption of active recruitment.

- The Agency must ensure the availability of critical skills, using appropriate incentives when necessary to recruit and retain employees.
- Recent downsizing and hiring limitations by the Agency may cause a future shortage of experienced leadership.
- The shortage of experienced, highly skilled workers has contributed to increases in workforce stress.

Unfortunately, NASA's need to reconstitute the workforce with the right skills and abilities is occurring at the very time in which competition for workers with those skills is intense.

Significant Loss of Knowledge Due to Potential Retirements Within the S&E Workforce

I have just discussed the skills imbalances that NASA faces today. The situation promises to worsen with time. New skills imbalances will occur over the next several years as the aging workforce reaches retirement eligibility. Approximately 15 percent of NASA's S&E employees are eligible to retire now. Within five years, almost 25 percent of the current workforce will be eligible to retire. Historical attrition patterns suggest that the percentage of those eligible for retirement should remain level at around 15–16 percent each year. In an agency where the expertise is not as deep as we would like it to be, even a few retirements can be critical. Everywhere I go across the NASA Centers, I hear the same story: "We're only one-deep. We can't afford to lose that skill." Clearly the Agency must begin preparing for its projected workforce needs now since a quarter of its senior engineers and scientists will depart this decade and the job market is far more competitive than in the past.

Another way to look at the potential loss of knowledge is to examine NASA's current S&E profile. At this time, within the S&E workforce, NASA's over-60 population outnumbers its under-30 population by nearly three to one. The age contrast is even more dramatic at some NASA Field Centers. While touring the Marshall Space Flight Center, I discovered that only 62 engineers out of a 3,000-person workforce were less than 30 years old. At the Glenn and Ames Research Centers the gap between the over-60 and under-30 population is five to one, and at Langley Research Center it increases to seven to one! By comparison, in 1993 the under-30 S&E workforce was nearly double the number of over-60 workers. This is an alarming trend that demands our immediate attention with decisive action if we are to preserve NASA's aeronautics and space capabilities.

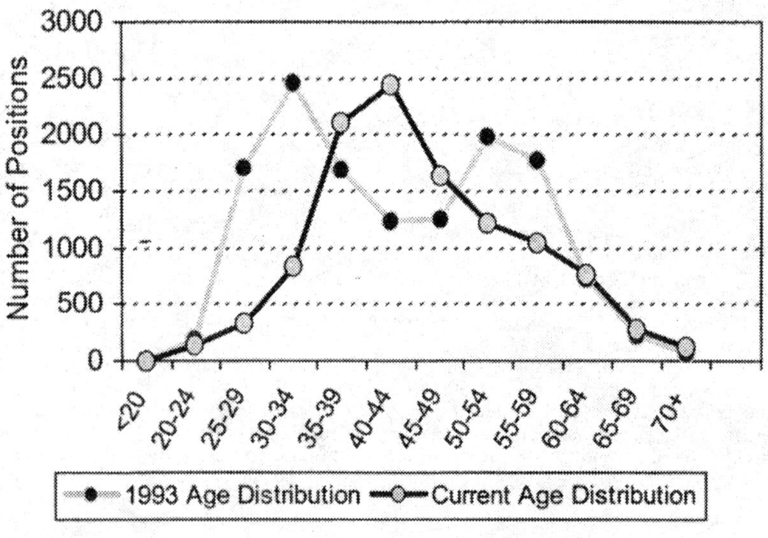

Increased Recruitment and Retention Problems

The last NASA trend I want to discuss with you today involves the evidence of increased difficulty of recruiting and retaining employees. Historically, NASA has enjoyed unusually low attrition rates, due in part to the attraction of our unique mission and the fact that our employees simply love their work and stay on the job longer than the typical worker. However, one recent trend is concerning. We have noted an alarming attrition pattern among NASA's most recent hires. Compared to an overall attrition rate of just under four percent for all S&E's, the departure rate for S&E's hired since 1993 is nearly double—despite the fact that in the fall of 2000 the Agency completed downsizing.

Another indicator of increased recruitment and retention challenges is the extent to which we at NASA have had to utilize monetary incentives to sustain a workforce with the right skills. The number of recruitment bonuses given to scientists and engineers increased significantly over the past three years. In FY 2001, 18 percent of the new S&E hires were given recruitment bonuses—in contrast to previous years in which the percentages were in single digits. In FY 2001 there were fewer S&E hires than in the previous year (521 versus 627), yet almost twice as many recruitment bonuses were needed to attract the desired recruits. Even utilizing all the tools at hand, we are at a disadvantage when competing with the private sector. The situation is particularly poignant at places like Kennedy and Johnson; the inability to entice their candidates of choice to work at the Cape or at "Mission Control" on the Space Program is a scenario few at these Centers could have anticipated.

Our challenge continues once we manage to hire our personnel. Although our historical attrition rates are low, we notice an alarming development among our youngest S&E population. After factoring out the 55+-retirement eligibility group, attrition among the S&E workforce is highest in the 25–39 age group. This phenomenon has a multi-faceted impact on NASA. It represents a lost investment for the Agency; shrinks the potential pool of future leaders and managers; and skews the average age of S&E workforce toward retirement eligibility age.

The need for retention bonuses to keep individuals with critical skills has increased dramatically as well. In FY 2001, we gave 38 such bonuses. This is greater than the *total number* of retention bonuses given for the previous four years!

Help Is Needed

All of these trends provide immediate warning signals that significant measures must be taken to address workforce imperatives that ultimately impact mission capability. We cannot resolve these new and emerging problems with past solutions, nor are current personnel flexibilities adequate.

To address the human capital challenges I have outlined for you today, NASA needs additional tools. We have used the ones we have and we have been innovative and imaginative but we need the Congress' assistance. Specifically, we need legislative solutions giving us the tools and flexibilities to:

- Encourage students to pursue careers in science and technology;
- Compete successfully with the private sector to attract and retain a world class workforce;
- Build a workforce that is representative of the Nation's diversity; and
- Reshape the workforce to address skills imbalances and gaps, and leverage outside expertise to address skills gaps and strengthen NASA's mission capability.

Each request in the legislative proposal has been carefully crafted to enhance NASA's ability to manage our human capital efficiently and effectively, in concert with the mandate of the President's Management Agenda—and plain old-fashioned good, sound management. Many of these provisions have been implemented by other agencies (such as the Department of Defense in their demonstration projects, and the Internal Revenue Service through their reform legislation), and are consistent with the Administration's Managerial Flexibility Act, now under consideration by the Congress. Without these legislative tools, NASA's challenges will soon become its crisis in human capital management.

Legislative Proposals

We are proposing several legislative provisions to address the threat to the S&E pipeline. Last summer, NASA submitted to the Congress the NASA Science and Technology Career Enhancement Act of 2001. This bill's *Scholarship for Service* program would offer college scholarships to students pursuing undergraduate and graduate degrees in science, engineering, mathematics, or technology. In return, the students would fulfill a service requirement with NASA following their graduation, thus providing a return on our investment. Current statutes do not allow a service obligation for scholarship recipients.

The *NASA Industry Exchange Program,* modeled on the very successful Intergovernmental Personnel Act authority, introduces a means for NASA to engage in mutually beneficial, collaborate ventures with industry to infuse new ideas and perspectives into the Agency, develop new skills within the workforce, and strengthen mission capabilities. Without such an authority, talented individuals from industry remain an untapped resource for the Agency since the salaries and benefits of the federal sector are not competitive with the compensation packages offered to industry's most talented workers. Assignments would be limited to two years, with a two-year extension, and would adhere to current statutes covering government ethics, conflicts of interest, and procurement integrity.

Enhancing the *Intergovernmental Personnel Act* authority to permit assignments up to six years (rather than four) is another tool that will facilitate knowledge transfer—an important goal of an agency that must sustain its intellectual capital. This flexibility will allow individuals from academia or other institutions to continue working in support of long-term projects or programs when the need for continuity is critical.

Several provisions of the bill enhance NASA's ability to recruit and retain a high quality, diverse workforce. The streamlined *hiring authority for severe shortage and critical needs positions* is necessary for NASA to conduct the hiring process more effectively. Using this authority, NASA could make "immediate" job offers to qualified individuals in shortage occupations or where there is a critical hiring need without the need to undergo the usual lengthy and cumbersome evaluation process. Eliminating delays in extending job offers in today's competitive market may mean the difference between hiring a quality engineer, scientist, or information technologist and losing the candidate to the private sector or academia.

The *category rating methodology* will provide NASA with a more streamlined candidate evaluation process that expedites the recruitment process, provides flexibility in selections, and preserves veterans preference and merit principles. The current method for evaluating and selecting candidates involves assigning candidates a point score between 70 and 100. The selecting official may select only from among the top three candidates. This is a cumbersome evaluation process that eliminates candidates based on numerical score differences, where such fine distinctions really are unrealistic and have little meaning. This complex process results in only a limited choice of candidates from which to select, and denies access to other potentially highly qualified individuals. With the proposed category rating system, tested with great success in other agencies, candidates would be evaluated using a simplified method that places them into general quality groups such as "highly qualified" and "qualified," rather than assigning a numerical rating. The manager could select anyone in the top group, unless there is a veteran. Veterans receive top selection priority. These provisions give managers wider latitude to select the individuals they find best suited to their jobs, while simplifying and accelerating the process for candidates and managers.

Enhanced recruitment, relocation, and retention bonuses will help us with enhanced authority to offer financial incentives to individuals to come to work for us, to relocate to take on a new assignment, or to remain with the Agency instead of leaving to pursue a more lucrative job opportunity or retiring. Current bonus authority offers up to 25 percent of basic pay, and has proved useful—to a point. Our proposal would base bonuses on the higher locality pay salaries, allow greater amounts when coupled with longer service agreements, and make more flexible payment options available (such as a choice between up front payments, installments, and payments at the conclusion of an assignment). These payment options could be tailored to the situation at hand, and tie payment of the incentive to actual performance.

The *term appointment authority* is used extensively within the Agency to support many NASA programs and projects. It is useful for work of a time-limited duration, and it allows the Agency to terminate employment without adverse action when the need for the work/competencies wanes. The bill's provision to allow a limited number of term appointments to be extended up to six years, rather than four, will enhance its usefulness by accommodating the length of some NASA programs and projects. In addition, the bill provides that a term employee may be converted to a permanent position in the same line of work without further competition, provided the employee was initially hired under a competitive process and the public notice specified the potential for conversion. This provision does not alter any feature or principle of the competitive process, but eliminates the need for duplicative competition. Ultimately it may make the concept of term appointments more attractive to potential applicants and thereby provide a more robust labor pool for NASA management to consider. Conversions of term employees to permanent positions that differ from the position for which the employee initially competed would require internal competition.

In order to attract world-class talent into NASA's most essential positions, we propose changes to the authority to pay employees in *critical positions.* We seek authority to grant critical pay for up to 10 positions per year, subject to approval by the NASA Administrator, with pay up to that of the Vice President (currently $192,600). These enhancements will help us compete in an enormously competitive job market. The provisions raising the annual *compensation cap for NASA excepted employees* appointed under the Space Act from Level IV of the Executive Schedule to Level III will address this need as well. Based on current salaries, this would allow an increase from $130,000 to $138,200.

Separation incentives *("buyouts')* and early retirements *("early outs")* are valuable tools to encourage voluntary attrition as the Agency rebalances skills and reshapes its workforce. NASA needs the tools to encourage targeted attrition in areas in which the need for certain skills has diminished so that it can recruit and reshape a workforce that is aligned to current and future mission needs.

Finally, the *streamlined demonstration authority* provision provides the Agency with an effective and extensively tested mechanism for pursuing additional human resources innovations in response to changing workforce needs. A number of agencies, notably the Department of Defense and Department of Agriculture, have operated highly successful projects. Unfortunately, the legal requirements to initiate a demonstration project are daunting. Current law limits "demo" projects to 5,000 employees, requires multiple *Federal Register* notices, a public hearing, and a 180-day notice to affected employees. Once an agency successfully tests a system, it must seek additional legislation to make that system permanent. The authority we are seeking would remove the coverage limit to allow widespread testing of new ideas, and shorten the steps to implement a project. Once a project proves successful, the Office of Personnel Management could approve conversion from a demo to a permanent *alternative personnel system* without further legislation.

Mr. Chairman and Members of the Subcommittee, each of these legislative provisions when taken individually will only help NASA deal with its human capital strategic threats to a limited degree. However, when taken together as an integrated package they form a strong nucleus in support of the Agency's Strategic Human Capital Plan and the President's Management Agenda, and will be invaluable as we deal with a diminishing pipeline, recruitment and retention of a world-class workforce, and skills imbalances. With these tools in hand, we will be able to avert a serious human capital crisis at NASA.

The missions we seek to lead and make possible are the visions that we all have for our future—new launch systems, innovations in high-performance computing, advances in biological research and exploration of our cosmos that extend our lives and way of life out there. Those things can only happen if we have the people that can make them happen. Technology and exploration will go nowhere without the human know-how and presence to make today's impossible into tomorrow's reality. After meeting and working with many of the men and women of NASA these past several months, I know we can do those things and I look forward to working with you and sharing the rewards of your investment and trust in us.

Biographical Data

National Aeronautics and
Space Administration

Washington, DC 20546
(202) 358-1600

NASA Administrator the Honorable Sean O'Keefe

Nominated by President George W. Bush and confirmed by the United States Senate, Sean O'Keefe was appointed by the President as the 10th Administrator of the National Aeronautics and Space Administration on December 21, 2001. As Administrator, O'Keefe leads the NASA team and manages its resources, as NASA seeks to advance exploration and discovery in aeronautics and space technologies.

O'Keefe joined the Bush Administration on inauguration day and served as the Deputy Director of the Office of Management and Budget until December 2001, overseeing the preparation, management and administration of the Federal budget and government wide-management initiatives across the Executive Branch.

Prior to joining the Bush Administration, O'Keefe was the Louis A. Bantle Professor of Business and Government Policy, an endowed chair at the Syracuse University Maxwell School of Citizenship and Public Affairs. He also served as the Director of National Security Studies, a partnership of Syracuse University and Johns Hopkins University, for delivery of executive education programs for senior military and civilian Department of Defense managers. Appointed to these positions in 1996, he was previously Professor of Business Administration and Assistant to the Senior Vice President for Research and Dean of the Graduate School at the Pennsylvania State University.

Appointed as the Secretary of the Navy in July 1992 by President George Bush, O'Keefe previously served as Comptroller and Chief Financial Officer of the Department of Defense since 1989. Before joining Defense Secretary Dick Cheney's Pentagon management team in these capacities, he served on the United States Senate Committee on Appropriations staff for eight years, and was Staff Director of the Defense Appropriations Subcommittee. His public service began in 1978 upon selection as a Presidential Management Intern.

Sean O'Keefe is a Fellow of the National Academy of Public Administration and has served as chair of an Academy panel on investigative practices. He was a Visiting Scholar at the Wolfson College of the University of Cambridge in the United Kingdom, a member of the Naval Postgraduate School's civil-military relations seminar team for emerging democracies and has conducted seminars for the Strategic Studies Group at Oxford University. He served on the national security panel to devise the 1988 Republican platform and was a member of the 1985 Kennedy School of Government program for national security executives at Harvard University.

In 1993, President Bush and Secretary Cheney presented him the Distinguished Public Service Award. He was also the recipient of the Department of the Navy's Public Service Award in December 2000. Sean O'Keefe was the 1999 faculty recipient of the Syracuse University Chancellor's Award for Public Service. He is the author of several journal articles, contributing author of "Keeping the Edge: Managing Defense for the Future," released in October 2000, and in 1998, co-authored "The Defense Industry in the Post-Cold War Era: Corporate Strategies and Public Policy Perspectives."

Sean O'Keefe earned his Bachelor of Arts in 1977 from Loyola University in New Orleans, Louisiana, and his Master of Public Administration degree in 1978 from The Maxwell School. His wife Laura and children Lindsey, Jonathan and Kevin, reside in northern Virginia.

April 2002

Chairman ROHRABACHER. Well, seeing that it took the White House so long to get a new Administrator in office, we felt we did have to move somewhat soon on some of your suggestions. Our final witness is Mark Roth, General Counsel for the American Federation of Government Employees. He will testify with some detail on NASA's legislative proposals before this committee and provide his perspective, which is important for us to take into consideration when looking at these reforms. So, Mr. Roth, thank you for being with us today, and you may proceed.

STATEMENT OF MR. MARK D. ROTH, GENERAL COUNSEL, AMERICAN FEDERATION OF GOVERNMENT EMPLOYEES, AFL–CIO

Mr. ROTH. Mr. Chairman, and Members of the Committee, I thank you for the opportunity to testify here today on the draft proposals for sweeping civil service exemptions at NASA. We represent a significant number of NASA's white collar employees, so we are very concerned. The bottom line in our opinion is that it is unwise to make necessary civil service changes on an agency by agency basis rather than government-wide.

The human capital crisis is government-wide. It involves scientists at Agriculture, EPA, FBI, CDC, and every other agency. The proposals contemplated in this legislation have already been presented elsewhere as government-wide changes and have been rejected largely on the grounds that they undermine merit system principles, that they would actually exacerbate the Federal Government's so-called human capital crisis, and that they would create serious conflicts of interest between private sector interests and the public good.

"HUMAN CAPITAL LEGISLATION REFERRAL TO GOVERNMENT REFORM COMMITTEES"

For these reasons and others, AFGE opposes most of the human resource proposals contained in this legislation. Further, it must be noted that the Committee on Government Reform has primary jurisdiction and expertise in matters involving Title 5. With all due respect, these proposals, some of which NASA admits would amend Title 5 government-wide do not belong in the Committee on Science with a jurisdiction and documented expertise in areas other than federal personnel.

For the record, AFGE opposes the approach again of seeking changes to civil service laws on an agency-by-agency basis. The Bill before you today is virtually a replica of the original version of the government-wide Freedom to Manage Act submitted by Senator Voinovich on OPM's behalf over a month ago to Senator Akaka's Senate subcommittee. We have literally spent dozens of hours working with the Senator and his staff in deleting objectionable provisions, refining vague or unacceptable flexibilities, and drafting what we believe are some acceptable substitute provisions.

We are not there yet, but I believe in addition to referring the Bill to the House Government Reform Committee, it would well serve the Committee to contact Senator Voinovich's staff for a copy of the latest version of the proposals. You have my full statement, and I would refer you to it. It goes into great length with respect to why we vehemently oppose NASA's proposals, objections that I cannot honestly detail in the depth they require in the five minutes I have been allotted by the Committee.

Human Capital Crisis

The Committee has asked what AFGE considers to be the most important human resource challenges facing NASA over the next 10 years. Like the rest of the Federal Government, NASA will experience a wave of retirements in that period as workers with between 25 and 35 years of federal experience reach retirement eligibility. Unfortunately, prior to the retirement wave and prior to the current Administrator's tenure, NASA has for more than a decade pursued a vigorous and ill-conceived program of downsizing and outsourcing. Instead of careful consideration of whether NASA's mission could be most effectively and economically carried out by hiring in-house employees, it has engaged in wholesale privatization, not competitive sourcing and made reliance on contractors the rule.

In effect, it is a contractor-run agency. This has proven to be a costly mistake for NASA both in terms of taxpayer dollars, and in terms of the Agency's internal human resource infrastructure, which we are trying to deal with today. Eliminating Federal positions and rushing to contract out as much government work as possible rather than building and planning for a transition to the next generation of NASA employees who are dedicated to career civil service with the Agency has made the coming retirement wave challenges truly daunting for NASA, and we sympathize with that.

AFGE Opinion on NASA's Employee Buyouts Proposal

But in this context the human capital proposals in the draft legislation are extraordinarily paradoxical. At the same time as they are offering buyouts to certain employees and getting them out the door, they are then imposing a limitation on filling that position.

At the same time, they are giving recruitment and retention bonuses, which far exceed current law, and so they are trying to bring in the door some people, but while they are targeting and taking people and trying to buy them out, they are putting quotas on and imposing restrictions on the FTE—filling those FTE positions when they perhaps should be encouraging senior employees to leave using those salaries to bring in one or two new junior people.

AFGE OPPOSES NASA'S IPA EXTENSION PROPOSAL

The idea—we oppose extending the Inter-Agency Personnel Act. This contemplates again people coming into NASA for short periods of time from private sector firms or for term appointments. We don't believe that a six-year appointment is a term appointment that cannot be made under current hiring proposals. We believe NASA should stay with current law there. It has the authority to do that.

AFGE OPINION ON NASA'S DIRECT HIRE PROPOSAL

Another provision that bothers us is the direct hire authority for NASA.

We question the necessity to evade these merit-based competitive procedures that are in place now. Logically direct hiring implies that agencies would create the opportunity to hire a candidate who would not have been eligible had he or she been forced to compete for the job, that is, by circumventing competitive procedures, agencies could hire someone other than the best qualified candidate. This is plain wrong.

AFGE OPINION ON EMPLOYEE BONUSES

The legislative draft also contains proposals for enhanced authority to offer, as I said, recruitment relocation and retention bonuses of up to 100 percent of salary over four years. We believe that the approach to financial incentives for recruitment and retention contained in this legislation is misplaced. Federal salaries are just plain too low, not just for prospective employees but for employees the agencies have currently on board.

The FEPCA Act has never been fully funded or implemented. Congress needs to do that. While we support bonuses, they are not a substitute for fully funding FEPCA. Also, there is current authority for bonuses for the payment of recruitment and retention bonuses. We supported it. We helped get it into law. But according to a comprehensive study published by OPM in 1999, less than one percent of eligible Federal employees had received bonuses. That is because they are not being funded.

When we look at the NASA legislation, they are not being funded here. That means that in order to give some people these bonuses other people will have to lose their jobs or somehow be reduced in grade. This will only exacerbate NASA's current woes. I would direct at this point the Committee to a hypothetical scenario I set out on pages 8 through 10 of my testimony for a good example of just how cockeyed NASA's flexibility provisions are. Suffice it to say that you will see that a recruited employee at NASA will be able to enter government on an expedited basis, leap-frog more qualified applicants, including veterans, then leave government after just six years of service having received more than 8½ years of salary. Nice windfall for the employee, bad news for current employees, veterans and taxpayers.

AFGE OPINION ON DEMONSTRATION PROJECT FOR PERSONNEL HIRING

One final point, and that is we do not oppose demonstration projects. We oppose their demonstration project proposal. We think they can do a lot of what they want to do under current Title 5, USC 4703, and we are willing to work with them on a demonstration project in NASA because when they bring in new employees, these are also new prospective members for AFGE. We have a lot of common ground, and we can do it under current law. Thank you very much.

[The prepared statement of Mr. Roth follows:]

PREPARED STATEMENT OF MARK D. ROTH

Mr. Chairman and Members of the Committee: My name is Mark D. Roth, and I am the General Counsel of the American Federation of Government Employees, AFL–CIO (AFGE). On behalf of the more than 600,000 federal employees our union represents, I thank you for the opportunity to testify here today on the draft proposals for sweeping civil service changes at NASA.

The Committee has asked me to address five broad questions in this testimony. They include AFGE's assessment of NASA's most important workforce-related challenges over the next ten years, our views of the most effective strategies for recruitment and retention, and whether the proposed legislation would further those goals, and finally, AFGE's own recommendations, and our opinion of the wisdom of seeking civil service changes on an agency-by-agency basis rather than government-wide.

The proposals contemplated in this legislation have been presented elsewhere as government-wide changes, and have been rejected largely on the grounds that they undermine merit system principles, that they would exacerbate the Federal Government's so-called "human capital" crisis, and that they would create serious conflicts of interests between private sector interests and the public good. For these reasons and others, AFGE opposes most of the human resources proposals contained in this legislation. Further, it must be noted that the Committee on Government Reform has primary jurisdiction and expertise in matters involving Title 5. With all due respect, these proposals do not belong in the Committee on Science where the jurisdiction and expertise is in areas other than federal personnel. AFGE opposes the implied policy of seeking changes to civil service laws on an agency-by-agency basis.

The Committee has asked what AFGE considers to be the most important human resource challenges facing NASA over the next ten years. Like the rest of the Federal Government, NASA will experience a wave of retirements in that period, as workers with between 25 and 35 years of federal experience reach retirement eligibility. Unfortunately, the policies both NASA and other federal agencies have been pursuing for the past decade are likely to exacerbate the problems and challenges the retirement wave presents. Prior to the retirement wave, NASA has for more than a decade pursued a vigorous and ill-conceived program of downsizing and outsourcing. Instead of careful consideration of whether NASA's mission could be most effectively and economically carried out by hiring in-house employees, it has engaged in wholesale privatization (NOT "competitive sourcing"), and made reliance on contractors the rule.

This has proven to be a costly mistake for NASA, both in terms of taxpayer dollars and in terms of the Agency's internal human resource infrastructure. Eliminating federal positions and rushing to contract out as much government work as possible, rather than building and planning for a transition to the next generation of NASA employees who were dedicated to career service with the agency has made the coming retirement-wave challenges truly daunting for NASA.

In this context, the human capital proposals in the draft legislation seem extraordinarily paradoxical. The proposals would encourage the elimination of even more federal positions, and thereby further undermine the Agency's ability to assemble a new generation of career civil servants dedicated to carrying out NASA's mission in the most efficient and reliable way for taxpayers. The proposals would establish NASA as a place where people might take a short turn through the revolving door between the agency and its contractors, but not a place to build a career, not a place that expects loyalty from its employees or that will exhibit any in return.

The first section of the Title concerning "human capital" proposes various expansions in NASA's authority for employee exchanges, including the establishment of an employee exchange program between private firms and NASA. While AFGE does not oppose allowing federal employees to have temporary "details" or assignments to private firms in order to gain insight or experience into a given firm's unique technology or organizational structure, we do oppose allowing representatives of private firms to spend up to six years inside a government agency.

We believe that allowing employees of private firms, whose loyalties and interests must always be to their employers, would have irresolvable conflicts of interests, as would the agency. NASA has a duty to taxpayers and the public to keep its contractors at arm's length. We believe that allowing private sector firms with an interest in NASA's operations to send its employees into the Agency for two to four years would allow them to "case the joint" for lucrative future contracting opportunities.

Once inside, the exchange employee would be in a privileged position to report back to his employer on what segments of NASA's operations would be most profitable for the firm to take over, and what should be left to the Agency. Additionally, the private firm's employee could gain access to financial and other information that would put the firm in a favorable position relative to not only the government's own employees, but also to other private firms that might someday compete for the opportunity to perform government work.

If NASA needs to hire individuals with private sector experience for positions which will last only from two to six years, it should do so under existing hiring authority. However, there is reason to believe that the "long term projects" that NASA envisions staffing through these "temporary" exchanges are neither temporary nor short-term, and the government as well as the employees deserve to have the security and continuity that a regular civil service appointment conveys. With regular permanent appointments, the government will have obtained the expertise it needs, as well as the authority to supervise, manage, and control that employee's time and work. Taxpayers' interests will be served by knowing that only federal employees, sworn to uphold the public good and work in the public interest, have access to privileged agency information and operations. For these reasons, AFGE opposes both the NASA-industry exchange program, as well as the extension of the Intergovernmental Personnel Act's maximum time period and its expansion government-wide.

The next section of the legislation proposes various new authorities allegedly aimed at facilitating the recruitment and retention of new federal employees. One section would amend Title 5 and therefore be applicable beyond NASA and include all agencies of the Federal Government. It would allow agencies to hire directly for Federal Government positions, without regard to existing competitive procedures. We question the necessity to evade these merit-based competitive procedures. Logically, direct hiring implies that agencies would create the opportunity to hire a candidate who would not have been eligible had he or she been forced to compete for the job. That is, by circumventing competitive procedures, agencies could hire someone other than the best qualified candidate. This raises obvious questions—what attributes make the candidate who could be hired directly but not competitively attractive to NASA or other agencies? If factors other than merit are permitted in the "awarding" of federal jobs, there is reason to believe that the merit system will be undermined.

The legislative draft also contains proposals for enhanced authority to offer recruitment, relocation and retention bonuses of up to 100 percent of salary over four years. AFGE is strongly in favor of NASA's willingness to focus on the importance of improving pay as a means of improving recruitment and retention. However, we believe that the approach to financial incentives for recruitment and retention contained in this legislation is at best incomplete, at worst, misplaced. Federal salaries are too low not just for prospective employees, or for employees the agencies expect to employ only for a short period. Salaries are too low for all employees. There are market-driven reasons why the Federal Government should pay competitive salaries, and there are values-driven reasons why the Federal Government should pay competitive salaries. While market-driven reasons such as recruitment and retention may on the surface only appear to apply to prospective employees and "flight risks," they in fact apply to all employees.

AFGE does support the use of bonuses and other financial incentives to reward federal employees. Yet they should never be used as substitutes for a fully funded regular pay system. The "human capital" crisis these bonuses are ostensibly meant to alleviate is in part a result of the repeated failure to implement and fund the Federal Employees Pay Comparability Act (FEPCA), passed in 1990. FEPCA already provides broad authority for the payment of recruitment and retention bonuses. According to a comprehensive study published by OPM in 1999, less than one percent of eligible federal employees had received bonuses under FEPCA's authority. The main reason for the failure to use existing authority cited repeatedly by agency managers was an absence of funding.

It is important to recognize that the draft legislation does not provide a separate, supplemental funding mechanism for the payment of bonuses. Implicitly, the assumption is that the bonuses would be financed from existing salary accounts. That is, the agency would only be able to use the broadened authority in the draft legislation if it paid for them through the elimination of jobs or the denial of other salary adjustments for those not selected for a bonus.

It is foolish to pretend that, if enacted, these provisions would improve NASA's ability to recruit and/or retain federal employees. Bonus payments do not count as basic pay for purposes of retirement or other salary adjustments. They are a poor substitute for the provision of competitive salaries and regular salary increases that allow employees to maintain decent living standards.

Before implementing a bonuses-for-some, instead of an adequate-salaries-for-all approach, NASA should ask itself the following: Should employees who are loyal and have made a decision to dedicate their careers to public service be penalized financially relative to those whose only loyalty is to their individual paycheck? Should the federal pay system reward only those willing to extort a bonus from an agency by continually threatening to leave in the middle of an important project? Or should the Federal Government pay adequate, competitive salaries to all its employees?

The draft legislation makes the following scenario possible: a recent college graduate is hired "directly" at a job fair, effectively beating out three other candidates who had applied for the position through normal competitive procedures (among the three were a veteran with relevant experience and the same degree from the same university, a disabled veteran with 10 years of federal employment and a similar degree, and a recent graduate from another university with the same type of degree but a higher GPA who mistakenly thought the best route to federal employment was to follow procedures and fill out a Standard Application Form 171). To encourage the direct hire person to accept the position, he is promised bonuses worth 25 percent of salary each year for four years (indeed, he must also accept a service agreement wherein he agrees to work for NASA for a period "not to exceed four years"). During that four-year period, the agency would repay the employee's student loans. At the end of the four-year service agreement, the employee indicates or threatens to leave in the middle of a project. NASA realizes a need to keep him on for at least two more years. A retention bonus of up 50 percent of salary, for two years, is authorized. At the end of this period, a new Administration/political appointee at NASA decides it can do without him, and offers him a Voluntary Separation Incentive Payment (VSIP), determined not on the basis of the regular severance formula in Title 5, but rather equal to 50 percent of his salary, since the Administrator has determined that to eliminate this position is critical to his restructuring efforts. Over six years, the lucky employee has received more than eight and a half years of salary. And the expertise and experience he has built up over that period is lost to the agency.

That's quite a windfall for the hypothetical employee, quite an expensive experiment for taxpayers, and quite an insult to the thousands of rank and file federal employees who are taken for granted and denied competitive salaries, benefits, or any form of job security. The question is: Is it a reasonable response to the impending "human capital" crisis? Will it allow the government to replace the more than 50 percent of federal employees who will be eligible to retire within the next five years with a new generation of employees who exhibit the same level of skill, dedication, and reliability as our nation has relied upon in the past?

Federal agencies, particularly science-dominated agencies like NASA are not fly-by-night operations or flashes-in-the pan. They are not here today and gone tomorrow, nor do they produce technological fads with only passing relevance or utility. As such, no federal agency, including NASA, should have a human resources plan that explicitly encourages constant turnover and puts no value on continuity, dedication, or career development for the incumbent workforce. Yet that is exactly the direction this draft legislation would take the agency, and in some cases, the entire Federal Government.

Additionally, in the context of the "human capital" crisis, granting NASA the authority not only to offer Voluntary Separation Incentive Payments (VSIP), and not only to inflate their size, but also to tie their use with a one-to-one elimination of full-time-equivalent positions is ridiculous. At a minimum, there should be no connection between efforts to restructure and delayer and authorized agency FTE levels. At a time when NASA and other federal agencies are asking for expanded authority to pay bonuses and repay student loans in their efforts to hire more federal workers, why should they simultaneously be required to eliminate FTE's and pay employees to leave federal service? Couldn't the money be better spent on retraining? On improving salaries? On improving the Federal Employees Health Benefits Program (FEHBP) which this year pays only 70 percent of the premium of the plan that covers over half of all enrollees?

Finally, the draft legislation calls for the establishment of authority in NASA for streamlined demonstration projects, and the ability to bypass Congress in creating permanent "Alternative Personnel Systems." It would also remove any limit on the number of workers covered by either the demonstration project, and reduce the time period for notice and comment by affected employees. Most dangerously, however, it would grant to the Office of Personnel Management (OPM) the power to make so-called "Alternative Personnel Systems (APS)" tried out in NASA demonstration projects permanent. Currently, only Congress has the power to make permanent changes to Title 5. AFGE strongly opposes this attempt to allow OPM to take over the role of Congress in this way.

Federal employees value their democratic rights to inform their elected representatives of their views regarding any proposed changes to Title 5. AFGE has always supported the use of demonstration projects to experiment with alternative personnel practices, including alternative pay systems. Yet the draft proposal, in attempting to give OPM authority that today rests solely with Congress, takes away any incentive federal employees might have to experiment with demonstration projects. Why experiment when the outcome could be a permanent elimination of Title 5 protections by OPM, with no opportunity for Congressional input? This proposal begs the question of whether NASA envisions an attempt to make permanent an "alternative personnel system" that could not gain Congressional approval. AFGE believes that if a demonstration project authorized under current law 5 U.S.C. 4703 is successful enough to merit permanent establishment as an alternative personnel system, then Congress is quite unlikely to stand in the way.

This concludes my testimony. I will be happy to answer any questions Members of the Committee may have.

Biography for Mark D. Roth

Mark D. Roth is the General Counsel of the American Federation of Government Employees, AFL–CIO. The "Dean" of Federal sector General Counsels, Mr. Roth started his tenure with AFGE in 1976 and, since 1983, he has served as AFGE's General Counsel under three AFGE National Presidents. In this capacity, he functions as the chief legal officer of the Union and supervises an in house staff of ten attorneys in civil litigation and administrative actions before FLRA, MSPB, and EEOC. In addition to being the counsel of record in numerous court appeals involving precedent-setting federal labor law issues, Mr. Roth was also counsel of record in *Frazier v. MSPB,* the landmark whistleblower case.

In 1991, Mr. Roth was one of the chief negotiators responsible for enacting collective bargaining rights for Title 38 employees in the "Physicians and Dentists' Compensation and Labor Relations Act of 1991." In October of 1994, he was a principal in the negotiations which, for the first time, provided Title 38 employees with specific whistleblower protections in the "Whistleblower Protection Reform Act of 1994."

As a member of the Board of Directors of the AFL–CIO's Lawyer's Coordinating Committee, Mr. Roth continues to shape labor law initiatives that affect the entire labor movement.

Discussion

Chairman ROHRABACHER. Well, thank you very much, Mr. Roth. Some of the things you were describing, they did—these were things that were started under the leadership of Mr. Dan Goldin, was it not?

Mr. ROTH. They were actually. I am going to go higher. They were actually Clinton Administration proposals. I gave testimony yesterday before the Volcker Commission on Public Service, and I shot it straight between the eyes.

Chairman ROHRABACHER. I just want to make sure everybody understands that.

Mr. ROTH. Everyone understand that this was—began with the wrong sizing under the Clinton Administration.

EXAMPLE OF NASA HIRING AUTHORITY

Chairman ROHRABACHER. Okay. Well, let us take a look at this example, one of the examples I gleaned out, and we will ask for some reaction from the Administrator and from the Administration here. You said about this fellow who comes in and after a short period of time leap-frogs everybody, but what if they find somebody in the private sector—and this is what I think it is aimed at, who really can make a contribution, who can maybe save the taxpayers millions of dollars, and may insure the success of a project—who maybe doesn't want to spend the rest of his life working for NASA, but has a major contribution to make. Doesn't it benefit us to be able to make sure he is attracted into that service and saves us that money and uses his unique expertise, and then is able to get out, realizing that it has been at least not a total sacrifice on his part?

Mr. ROTH. Well, first of all, I believe they can probably hire that person under current hiring authority, and they could leave at their choice after six years, but also I think you have to look at the reality of the NASA workforce. It is a contractor workforce right now. It is overwhelming—the work done is overwhelmingly being done by contractors, and bringing this person in for six years as a quasi-federal employee does not get it done. They then leave with that institutional——

Chairman ROHRABACHER. Okay. I got your answer. Mr. O'Keefe.

Mr. O'KEEFE. Thank you, Mr. Chairman. First of all, I think the hypothetical that you proposed is certainly more appealing than the limitations I have heard here. Mr. Roth, I've just got to respectfully disagree. I think there is a distortion of the example you used. It does not supersede veterans preference under any circumstances. The difference is that there is a period of time that it takes and that averages at NASA about five months, in other agencies it is as little as three, that this approach would bring down to about 15 days to select qualified candidates from the most qualified banding for which veterans preference would supersede all other factors. So I really have to disagree with you on the implications in that circumstance because it is a very important one, and it is one we really sought to preserve by virtue of significance.

Chairman ROHRABACHER. Well, he even gave a specific leap-frog example. Let us take a look at it.

Mr. O'KEEFE. And that permits mid-level entry at an approach where you can recruit someone from other than fresh out of undergraduate school or graduate school or any other area from industry, potentially from a variety of university activities in science and engineering approaches that would have to be in that band of very well qualified in order to be selected under this direct hire authority.

Chairman ROHRABACHER. Well, he says you can already do it. We don't need to have any changes for that.

IPA Term Extension

Mr. O'Keefe. We have very limited authority right now. We have under the Intergovernmental Personnel Act authority today that we can utilize, and the current limitation is it must be renewed every two years.

Chairman Rohrabacher. Say this again now.

Mr. O'Keefe. For every two. It is a two-year appointment that will last for no more than I believe the maximum right now is four, and that is from universities or a variety of other non-governmental——

Chairman Rohrabacher. So you would like to see those restrictions on every two years lifted.

Mr. O'Keefe. I would like to expand it to six.

Chairman Rohrabacher. Expand it to six.

Mr. O'Keefe. And have an opportunity to broaden it to include an Intergovernmental Personnel Act that would permit from universities as well as other agencies the opportunity to look at a very specific patterned program after exactly the same authorities for those areas looking at business sector kind of backgrounds.

Outsourcing NASA Work

Chairman Rohrabacher. All right. We will leave it to people to determine whether they think that is a reasonable request. Let me just in a general way to Mr. Walker, what about it? I mean, is contracting out better than hiring an employee or this management expert here? And we have heard that we are not going to get the top quality employees to sign on with the government, if we end contracting out this way.

Mr. Walker. Well, first I think it is important to note that NASA contract management is on GAO's high risk list, and there is a good reason it is on GAO's high risk list. Because irrespective of the merits of whether or not you contract out something, and I'll come back to that, you have to be able to have an adequate number of qualified parties to be able to manage the contractors, as to cost, quality, and performance. And if you don't have that, you are going to get in trouble.

As far as what should be contracted out, as you know, Mr. Chairman, I recently had the opportunity to chair something called a Commercial Activities Panel at the request of Congress. And that involved labor leaders, including Bob Harnage, administrative officials, some people from academia and others. We came out with a series of recommendations, which came before the Congress in May and are now being considered as to what are the criteria that should be considered in determining sourcing decisions. Sourcing meaning not just outsourcing, but insourcing and cosourcing, and I would commend that to the Committee.

GAO Opinion of Term Appointments

I think frankly, Mr. Chairman, there are opportunities to look at things that Congress has already enacted as possibly being a model for reasonable flexibility with appropriate safeguards to prevent abuse. Some examples that I would give would be the GAO. The ability to make term appointments for a stated period of time. At the GAO, it's three years. I can appoint up to a certain number of people at any level for a specific task for up to three years, without going through the competitive process. But there is a limited number, as there should be, that you can do.

GAO OPINION OF REALIGNMENT AUTHORITY

Secondly, realignment authority, not to downsize the Agency. A realignment authority that would say that you can offer early outs and buyouts to be able to help realign the Agency and position it for the future, rather than to downsize the Agency. Having authority to have scientific and technical career tracks, having the ability to pay additional compensation for critical occupations, having the ability to potentially allow people to gauge in phased retirement.

Chairman ROHRABACHER. Okay. Now these are all—the question——

Mr. WALKER. And these are the things that I think that is going to be——

Chairman ROHRABACHER. Well, the question I think was about contracting out, and let me suggest that we need some very high quality employees that are full-time employees that understand that they have long-term loyalty to the Agency and to the goals in mind for the Agency. But I do not think that what we also need is to have a situation dominated by people and totally dependent on people who just consider themselves to be long-term employees who are guaranteed their job and they have much more guarantees than they would have in other jobs in the private sector. Yes, Mr. Roth?

Mr. ROTH. Yes. I just want to make one point on that.

Chairman ROHRABACHER. Sure.

NASA CONTRACTOR VERSUS GOVERNMENT WORKERS

Mr. ROTH. Again, I have already said NASA is pretty much a contractor agency. But what I would challenge the current Administrator, and admittedly long before his tenure, this began. He can probably come back tomorrow and give you the name, classification, pay grade and number of every Federal employee employed at NASA. If you ask him to do the same for every contractor employee, you could be here a decade. I challenge him. I do not believe he can tell you the numbers of contractor employees and what their grades pay and other things they are doing.

Chairman ROHRABACHER. Of course, the object is not to increase the number of federal employees or people who are employed by the government, but to get the job done.

Mr. ROTH. But in order to manage, you have got to manage this other workforce, and I am saying I challenge him to say that that is being done.

Chairman ROHRABACHER. Well, I would—this is just my perspective, Mr. Roth, is that yes, there are certain benefits who are professional and on the job and have job security. I do not believe necessarily that we could look to government as an example of better management, of how to utilize employees' times necessarily. However, there has been some recent examples in the private sector where management at the highest levels have not met those expectations. We hope Mr. O'Keefe of course gives us a much better example of that than some of his colleagues in the private sector. Mr. Gordon, go right ahead.

NASA DOWNSIZING AND PRIVATIZATION

Mr. GORDON. Thank you. Administrator O'Keefe, the legislative provisions submitted by the Administration seem geared to enable more downsizing and more shifting of NASA's activities to the private sector. For example, NASA's documentation says that the proposed Section 203 on NASA industry exchange can, I quote, "Be used to assist employees whose jobs transfer to the private sector in connection with privatization actions." And as written, there are no limits on the number of employees who could be involved. Is that correct?

Mr. O'KEEFE. No, sir. I am not aware that there would be no limit of those employees involved. But I——

Mr. GORDON. But it is written—that is the way it appears to us. Could you clarify? I mean, I am not saying you are supposed to know this by heart right now. But if you could just get back to us and clarify that please.

Mr. O'KEEFE. I would be delighted to do that.

NASA'S BUYOUT PROPOSAL

Mr. GORDON. Another example is the proposed Section 209 on buyout authority. As written, it requires the total number of NASA employee provisions to be permanently cut by the number of employees who take the buyout option unless the OMB Director decides to intervene. Is that correct?

Mr. O'KEEFE. Well, the intent of the provision is to specifically eliminate a linkage between FTE reduction and the number of buyouts you use. The idea is to use it for workforce shaping. So, the objective is to say, you know, you want to hire one individual for one set of activities that may be totally unrelated to those that are at retirement eligibility or approaching it for which there is an abundance of talent in that particular field.

Mr. GORDON. Sure. But that's a more reasonable approach than the way the provision seems to be written. Would you review that also?

Mr. O'KEEFE. Certainly. That's why we have hearings, to try to get intent to the written statements so that we don't wind up with something that we—that you don't intend.

NASA'S ALTERNATE PERSONNEL SYSTEM PROPOSAL

Mr. GORDON. Finally, the proposed Section 211 would allow OPM to completely and permanently change NASA's personnel system, and if it chooses, change workforce provisions that are currently part of Title V Civil Service Law. As I read the provisions, NASA would be able to propose a personnel system that change rules on such things as sick leave, overtime, firing procedures, compensation, and employee rights of representation at any kind of hearing, all without any prior Congressional approval being required. Is that correct?

Mr. O'KEEFE. Yeah. The pattern of exactly that language was enacted by Congress a year ago to the Department of Defense and is being implemented today by the Army Medical Command.

Mr. GORDON. So, that is correct then?

Mr. O'KEEFE. Yes, sir. It is.

Mr. GORDON. So, your intention is to allow NASA to unilaterally propose personnel system changes that include sick leave, overtime, firing procedures, all of those? That's what—that's your intention?

Mr. O'KEEFE. The pattern of the precise marketing procedures we have seen previous pilot approaches enacted by Congress——

Mr. GORDON. Sir—I'm sorry, sir. Pardon me.

Mr. O'KEEFE [continuing]. For exactly the same purposes.

Mr. GORDON. Okay. You said that. Do you envision changing the rules on sick leave, overtime, or firing procedures?

Mr. O'KEEFE. I can't imagine that. No, sir.

Mr. GORDON. Then why would you want that authority?

Mr. O'KEEFE. Again, there is a series of piloted programs that have been enacted effectively.

Mr. GORDON. But I'm talking about in your agency, you know. Why would you want that authority if you're not intending on using it?

Mr. O'KEEFE. It is the intent of looking at an alternative personnel system that has been demonstrated across the government. We have an opportunity to look at the best practices to utilize that approach.

Mr. GORDON. But do you intend to change any rules on sick leave, overtime, or firing policies?

Mr. O'KEEFE. I think you asked that question.

Mr. GORDON. Right. Now, you haven't answered my question.

Mr. O'KEEFE. I apologize.

Mr. GORDON. I'm asking it again hoping that I'll get an answer.

Mr. O'KEEFE. I apologize. The intention is not to do that.

Mr. GORDON. Okay. That's all. Thank you.

Mr. O'KEEFE. Thank you.

Mr. GORDON. So, if that's—your intention is not to do it, then I would hope that you wouldn't ask for that authority until you better explain to us in the SRR where you intend to go.

Mr. O'KEEFE. Well, the intent was not specifically to cover the issues dealing with sick leave, annual leave, or you know, the. . .

Mr. GORDON. Should we take those out then, since that's not your intention?

Mr. O'KEEFE. Well, sir, again I'm going to ask your indulgence to permit me to say once again——

Mr. GORDON. Okay. Sure.

Mr. O'KEEFE. Thank you, sir. The intent is to utilize the best practices for which previous authorities or previous agencies of the Federal Government have been granted this alternative personnel system approach, so we can devise that to the extent we see what those best practices would be.

Mr. GORDON. All right. But you don't intend to use though. You don't intend to change the rules on sick leave, overtime, or firing procedures?

Mr. O'KEEFE. I don't believe that to be the case. No. I don't see——

Mr. GORDON. Okay. But——

Mr. O'KEEFE. And I don't know that to be a best practice deployed by anyone else either.

Mr. GORDON. Well, then again, it would be my opinion that if you don't intend to use it, then maybe you don't need that authority until you better explain to us where you're going and what you're going to do. But let's move on.

NASA-SPECIFIC GOVERNMENT-WIDE LEGISLATIVE PROPOSALS

Mr. O'Keefe, if I look at the Administration's proposed NASA provisions, I'm struck by the fact that a number of them would change workforce policies government-wide, not just at NASA. For example, one of the provisions would change the rules on IPAs, not just at NASA, but throughout the Federal Government. Another provision would change the rules that govern term appointments, not just at NASA, but throughout the Federal Government. Was that your intention?

Mr. O'KEEFE. It is identical to the provisions of the Managerial Flexibilities Act which was submitted last October.

Mr. GORDON. So, it is your intention then to change these Acts government-wide and not just within NASA?

Mr. O'KEEFE. Consistent with the Administration's proposal last October. Yes, sir.

Mr. GORDON. So—but that's what you intend to do?

Mr. O'KEEFE. That's right.

Mr. GORDON. Okay. So, have you talked to Secretary Mineta, Secretary Whitman, or Attorney General Ashcroft? Is that what they want also?

Mr. O'KEEFE. Yes. That provision was proposed by the President in the Managerial Flexibilities Act last October. And all of those agencies concurred in that position.

Mr. GORDON. And are you interested in those agencies proposing changes that will affect NASA through their own authorizations?

Mr. O'KEEFE. Any other provision of the President's Managerial Flexibilities Act that was submitted last October will be delighted to see included for our application. Yes, sir.

Mr. GORDON. But it just seems to me, you know, contrary to your testimony. In your testimony you said you wanted to take one agency at a time. Yet, here you are asking us—you are asking NASA to change policy for the whole government.

Mr. O'KEEFE. Well, sir, there are a couple of provisions that I think you rightly point out that do have peculiar application to NASA that we do seek your indulgence and consideration of those for those purposes. They are—because of the unique circumstances we confront given the science and engineering intensity of the nature of the workforce and so forth, we would just offer it for your consideration in that case.

NASA WORKFORCE DOWNSIZING

Mr. GORDON. Mr. O'Keefe, you have indicated the importance of acting on these provisions as a package. I can only assume that the Administration wouldn't have sent them over unless you have a clear plan for where you want to take the Agency. I think you need to share that plan with Congress and with the NASA workforce. For example, NASA's workforce has been cut from 25,000 to 18,000 over the last decade. How many more jobs do you envision being eliminated in the Agency and what is your target?

Mr. O'KEEFE. I don't have a target. And I would not speculate on how many would be eliminated or added at this time.

Mr. GORDON. But when would you let us know that?

Mr. O'KEEFE. As part of our strategic management of human capital plan that was just submitted and prepared, I'd be happy to share with you all the details of that objective. It does not prescribe targets. It does not envision very specific——

Mr. GORDON. But when would you—in some way you have to make those decisions. When would you expect making those decisions?

Mr. O'KEEFE. In each of the affected areas that we are working with across the Agency, our intent is to apply the strategic management of human capital planning that we put together to shape the workforce in that manner.

Mr. GORDON. And when would you expect—well, just take one, for example. When would you expect to let us know about that?

Mr. O'KEEFE. Well, if I could borrow from the Comptroller General's comment, this journey that is going to take us the better part of five to seven years to shape the workforce is where we intend to start now. And that is what is laid out in this plan.

Mr. GORDON. So, do you intend to let us know that you are going to take the workforce down, say, from 18 to 16 before you do it?

Mr. O'KEEFE. Well, yes, sir. Of course But I don't have any—please let me correct the undercurrent of your question. I have no intention, nor do I envision a specific increase, a target, or a decrease that would be involved.

Mr. GORDON. And so, it just seems you are asking for a lot of authority without giving us, you know, what we're trying to do with some oversight. And so, to give this authority blindfolded makes me a little bit nervous.

Mr. O'KEEFE. This is a chicken and egg question and an issue that I understand entirely what your concern is. The issue is do you provide the authority and then develop how it is implemented for the Congress's consideration with each budget submission for which you have the ultimate authority to determine whether or not that be enacted in that approach, or not, utilizing those specific authorities or do we propose some specific set of plans that pertain to individual programs that assume the enactment by Congress of a more general authority? And so, we have chosen to take the approach and propose for your consideration the implementation of policies which could be implemented to achieve a multiple set of effects on programs——

Mr. GORDON. And——

Mr. O'KEEFE. I am sorry. It will then be for your consideration as to whether you approve it, sir.

Mr. GORDON. I will close it. And I understand, as you say, that chicken and egg. But I think you are going to take—I am not sure whether you are the chicken or the egg approach. But whichever one you are taking, we would feel more comfortable with more information. And I think that is what we need. If you are going to say trust us, then give us a better idea of where you are going to go and what you are going to do.

Mr. O'KEEFE. Well, I don't believe there is a whole lot of trusting in the proposition we put before you. It is more a case that to the extent we implement the specific authorities, you will see it evident in the budget proposals before you and you will have an opportunity in Congress to make a determination as to whether those are to be approved or not.

Chairman ROHRABACHER. Thank you very much. I would note it sounds like the goal is flexibility. And when your goal is flexibility rather than a specific end, it is very difficult to set some solid parameters because your goal is not to have solid parameters, but instead to provide maximum flexibility. Now, I understand where Mr. Roth, of course, would, representing the employees, would not want someone else to have the flexibility because then he is looking for the rights of the employees. And there are two things—you know, there are different dynamics at work here. And I believe we are done with the questions. And I think we could have a second round and provide everyone with equal opportunity to go in depth, as Mr. Gordon was just able to do. I would now turn to Dr. Weldon.

OVERALL INTENT OF NASA'S HUMAN CAPITAL PROPOSALS

Mr. WELDON. Thank you, Mr. Chairman. Mr. Roth, if I understand it, the intent of the proposal being put forward by the Administration is to make working for government agencies more attractive for young people coming out of college, for people who are currently in a university position or an industry position to move over. And specifically the things that they are looking for are basically that they are—that we are competing with an industry that has performance-based pay with rapid advancement. That is what a young person is attracted to today. It is not the current hallmark feature of the Civil Service System, which is job security. And some people would criticize the current system as job security disconnected from work performance. What they are looking for is an opportunity to advance rapidly in a competitive environment. And that the current Civil Service System is just not competitive. How do you respond to that? You know, as I see it, what the Administration and Mr. O'Keefe is trying to put forward is to make NASA an attractive place for people to come to and an attractive place for people to stay once they come there, but most importantly to attract the brightest and the best. And the way the system is currently structured, we are not competitive in government. We are not competitive specifically in NASA.

Mr. ROTH. Well, generally let me say I do support the scholarship program that they proposed. We also believe it should be applied to current employees so they can upgrade their skills. The only problem we have with it is, again, it has no real guidelines. It requires Agency approval. But that is, I think, a fairly simple task, saying what type of degree programs you would fund, what type of scholarships you would have. We don't have a major problem with that. Quite frankly, we don't have a major problem with the general term "flexibility." We would like it further defined. Where we have been made part of the pre-deliberative processes in many, many agencies, the Mint, Social Security, Veterans Affairs Department, Department of Defense, in a lot of the demo projects he is talking about, we have the ability to come to the table in a non-bargaining context and actually work out compromises. Just—you weren't here when I said——

Mr. WELDON. So, what you are saying to me—the way I read what you are saying is that where this has been tested, the response of employees has been positive. Is that correct?

Mr. ROTH. It has been across the board. What I am saying is where there has been serious union involvement there have been compromises made. The flexibility—the Freedom to Manage Act, I call it the Freedom to Mismanage Act. As I said before. . .

Mr. WELDON. I want to interrupt you. I really take exception to that. Okay. I don't think these people have anything other than the best intention for the Federal workforce. And why are you implying in front of this committee—I mean, he has got a serious—I represent Kennedy Space Center. We have a serious human capital problem at Kennedy Space Center.

Mr. ROTH. No question.

Mr. WELDON. A third of the employees are eligible for retirement right now. The only reason they stay is they love their work. And what I am told is the young people that are coming out of college and going to work there, the only reason they are going there is they are enamored by working for the Space Program. They are offered much better jobs out there in the private sector. And we have the Administration coming forward trying to remodel our Civil Service System to make it more attractive for these people to come in. And you come before this committee and you imply that they are just trying to screw the workforce, and that is just not the case.

Mr. ROTH. I don't think I implied that. I said that this Act is schizophrenic. It has inconsistent approaches. And before you came, what I said was we were working with Senator Voinovich who introduced it in the Senate to make a whole lot of compromises and come to some reasonable approaches in such areas as VISUPS and VERAS and try and get it so that they are not then tied to FTE ceilings which they are in this Bill, if you read it. If someone takes an incentive, you cannot then fill the position. That is not a restructuring of the Agency. That is not a long-term solution. I share your views. I share your goals. AFGE does. We want a long-term solution to this. And what I——

Mr. WELDON. Okay. I want to work with the AFGE. I am the Chairman of the Civil Service Committee. You know that. And I would sincerely like to work with you. I just want to take one issue with one other thing that you mentioned. Okay?

CONTRACTING OUT FOR NASA'S WORK

And I would like to give you some time to elaborate on this a little bit. You said some of the contracting has been a costly mistake.

Okay. And I would like you to elaborate on that a little bit, certainly as it pertains to NASA because we went through a very difficult thing with that as far as the Space Shuttle Program was concerned. We took that which was primarily managed by civil servants, and we have essentially turned most of it over to the contractor which is the United Space Alliance. In the process, and it occurred under the Clinton Administration——

Mr. ROTH. Definitely.

Mr. WELDON [continuing]. We lost 2,000 employees at Kennedy Space Center. A billion dollars has been taken out of the Shuttle Budget, a very difficult process that I had a lot of concerns about. But in the end, the shuttle is operating more efficiently. Launches are going off more often on time. All of the objective measures are it's better. It's not worse. So, maybe it is other NASA Centers where the contracting has been a costly mistake. But from my perspective even though I had a lot of anxiety about it and objections to it as we worked all through it, ultimately in the end I have had to confess to Dan Goldin and Sean O'Keefe that you guys were right. You are operating the Shuttle Program more efficiently at reduced costs to the taxpayers.

Mr. ROTH. There are some areas where contracting out makes sense. AFGE is on record as saying that we support competitive outsourcing where there is a competitive process, an in-house bid versus a public bid. In a lot of areas that has not occurred. It has been single source or otherwise. There have been well-documented cases where contractors have taken some shortcuts on parts of components and those have not worked. I think you know them probably better than I. I'm a personnel analyst. But they have been all over the news in the last decade. And again, if I—I'm hopefully not saying this as Republican versus Democrat. This started under the Clinton Administration. I couldn't be more clear. I tried to say to the Chairman there was untargeted downsizing. If Mr. O'Keefe here today had been on board, perhaps he would have looked at it differently and said these positions can go and these other positions can not go and there would be a different composition of the workforce. But that didn't happen. But this is government-wide. This human capital crisis is government-wide. I know your concern is NASA. But our concern is also the other 50 agencies where we have employees and people are ready to go——

Mr. WELDON. My time has expired.

Mr. ROTH [continuing]. Center for Disease Control, FBI, the engineers throughout government. They are all seriously underpaid. I don't see any new funding in this program. That is all I was saying.

Mr. WELDON. Well, thank you. My time has expired. I just want to underscore that I am concerned about the government-wide issue as well. But NASA is my focus because of my district. But I share your concern about the government-wide issues as well.

Mr. ROTH. And I expect we will work with you in the other committee.

Chairman ROHRABACHER. And it is a pleasure to take the prerogative of being the Chair. And just since Mr. Walker was on the edge of his seat and wanted to say something, so maybe you would like. . .

GOVERNMENT-WIDE AND NASA HUMAN CAPITAL ISSUES

Mr. WALKER. I think I might be able to help provide some context which I think is important. I think what we have here is a macro issue, which is government-wide, and a micro issue which is NASA. On a macro basis if we look at our current human capital policies and practices in the Federal Government, it is based on the Industrial Age. It is based upon a government workforce that existed in the 40's and 50's. We are in the Knowledge Age now. Most of the decisions in government now are based on passage of time, rate of inflation, geographic location. We should keep Veteran's Preference. We should keep some of these things. No doubt about it. We need to move to a system that is focused more on skills, knowledge, and performance. Now, you obviously are confronted with NASA. There are opportunities, if you so desire, to provide NASA with some reasonable flexibilities, with more definition, with safeguards to have a better sense as to what they are doing and what they are not doing with periodic reporting, with periodic oversight, and with looking at it as part of their budget submission. You have done that. Congress has done that for certain agencies. And we are happy to help provide additional information if it would be helpful, Mr. Chairman.

Chairman ROHRABACHER. Thank you very much. Let me note one factor here. We do have a situation that in the private sector people can go on strike, and they are able to watch out for their own interests and their own rights by going on strike. We don't expect Federal employees to go on strike. So it is important that we have this type of dialogue to make sure that while we are trying to restructure and make things better, that the rights of our Federal employees are respected and we are doing what is fair by them as well as what is right by the taxpayers of the country. Ms. Jackson Lee, would you like to proceed?

NASA VISION AND PROGRAMS

Ms. JACKSON LEE. It is almost a question of where to proceed and how to start. Let me thank again the witnesses for their presentation. I do think for the record it is important to give, if you will, thanks to the visionaries that started with a president called Camelot who began this journey with us in this country and had so many of us with inspiration in our hearts and in our minds as we watched Americans brave as they were, some called, the right stuff going into space. And I guess my frustration is that we have come to a point where we seem to ignore either those courageous souls or those victories that we have had. I asked the staff to share with me whether alongside of these legislative proposals we have a vision. And certainly there is a vision statement that sounds extremely eloquent and poetic. It is a blueprint for the future of exploration and a roadmap for achievement that we hope will improve the lives of everyone in this country and everyone on this planet. Our new vision is to improve life here, to extend life there, and to find life beyond. But even in those very weighty words, there is no direct, concrete, visional, or visual program that tells us what we are going to do. I would suggest to the Administrator that all change is not progress. However, I don't think all change is bad. And I am willing to listen and to learn. But if I might cite again the fact that NASA seems, over the years, and might I share this as a bipartisan perspective, the previous Administrator was through a number of Administrations, Republicans and Democrats. And so, my efforts, of course, are to look at NASA as an entity for America, not a Democratic entity or a Republican entity. Well, NASA has consistently commissioned studies and reviews only to ignore them. And as I said earlier, the National Academy of Science Review Panel commissioned by NASA ranked a mission to Pluto, to Europa, and to other areas of interest as the highest scientific priority. Yet, this mission was eliminated from the fiscal year 2000's re-budget request. A NASA administrator stated that he will only fund a reduced crew for a complete space station although an independent task force again commissioned by NASA concluded that such a slimmed down force would not even be capable of carrying out the high priority list of research projects on the station. And most frightening for our young astronauts although the Administration repeatedly urged to develop a U.S. crew return vehicle as soon as practical. The NASA administrator has canceled X–38/CRV program without identifying a viable alternative for rescuing crew in the event of an emergency. My concern is that as we speak in high tones about recruiting and providing new talent, what are we giving them? What is their incentive to even come here? I am grateful that there are those in this audience that are in a particular program. I hope those programs will not be unique to one area. As I said before, I have been attempting to do one in our area. And I hope NASA will respond, as they have not done previously to my request. In addition, let me say to you that I have heard from a number of African-American women astronauts who are consistently sitting, gathering dust and mold, who have not been given the opportunity to do their best. It is my understanding that Mae Jemison, more than a decade ago, I believe, was the last African-American woman to fly into space. I will stand corrected if someone has the information. I would like a response to that. I would also like to have a response, a direct response, because I believe that the collaboration between private and public, Mr. Roth, has been good in many instances. There are in my community, positive results. But what I would like to know for a factual basis, what are the numbers that you expect to cut from each center? Right now in the NASA family you have people cowering in the corners and shaking in their boots that you are coming to slash them at any moment. And this kind of fuzzy talk is not going to help us understand who are you going to cut. What is the plan to cut personnel who believe that you are no longer committed to the vision of the Shuttle Program, the International Space Station Program? And let me just simply say, if I can get answers from Mr. O'Keefe quickly and Mr. Roth, Section 203 and Section 204, Mr. Roth, if you can comment on it, deals with the NASA Industry Exchange Program and the Direct Hiring Program, how deadly is that as it relates to those longstanding employees that are there who need incentives? And then, Mr. O'Keefe, you've heard me talk about the status of African-American astronauts and the question of slashing personnel and what you're doing about that. But then the bottom line is why should anyone, whether you have bonuses, or retention, or recoupment opportunities, even want to come to NASA? We really don't know what we are doing and what our mission is all about.

Chairman ROHRABACHER. Ms. Jackson Lee, you have asked some very provocative questions. And I think we should give our panel a chance to answer.

Ms. JACKSON LEE. And I am doing that, Mr. Chairman. I am yielding to them at this point.

Chairman ROHRABACHER. Okay. Thank you.

Ms. JACKSON LEE. Thank you very much.

Mr. O'KEEFE. Let's proceed, Mr. Chairman. There was two questions directed. Can I begin?

Chairman ROHRABACHER. Ms. Jackson Lee has asked a number of questions, and you may proceed to try to answer those questions.

Mr. O'KEEFE. Thank you very much. First of all, Councilwoman, I was not aware of the request you had pending. But I guarantee you I will go back immediately after these hearings and inventory where those requests are and be sure you get responses to them. I apologize.

Ms. JACKSON LEE. I thank you very much, Mr. Administrator.

Mr. O'KEEFE. I am advised that Mae Jemison was, in fact, the last female African-American to fly. There are currently three African-American female astronauts in the astronaut corps awaiting assignment at present. I will determine from Charlie Precourt, the head of the astronaut office, as soon as we leave here as well, at what point——

Ms. JACKSON LEE. Thank you. Mr. Chairman, I would just like to say I have been treated very poorly. But thank you very much. I look forward to that answer.

OUTER PLANETS EXPLORATION

Mr. O'KEEFE. I'm sorry. I will inquire as to the status of when their assignments will occur and see if we have a determination on that front as well. So I thank you again for raising the question. And let me just, I guess, select two of the topics that you have covered given the time involved to give you a sense of what kind of exciting programs that we are looking to that we hope will, in fact, be the excitement and the enthusiasm that we'll continue to want. Folks will want to be part of this stored institution. The first one that you mentioned, the Outer Planet Study was just released about 10 days ago. It did, in fact, rate Pluto as one of the mid-range missions that should be the highest on the priority list for mid-range missions, as in mid-expense. We are looking at that very specifically in terms of what the impact is, because today our circumstances that we would in order to accomplish that objective have to launch a probe by the year 2006, which would not even arrive at Pluto until the year 2018.

We have to find a way, therefore, to improve power generation and propulsion capacity. So what we are doing, and it is in the budget proposed for you in fiscal year 2003, is to aggressively begin a power generation and propulsion program that can cut down the distance and time it would take to achieve those kind of outer-planet objectives that, again, the National Academy of Sciences has very thoughtfully given to us as a range of exploration opportunities that should be pursued——

Ms. JACKSON LEE. You failed——

Mr. O'KEEFE. I'm sorry?

Ms. JACKSON LEE. Have you cut the mission out for 2003? You've been transferring the mission concept to just some research dealing with power generation. You cut the mission.

Mr. O'KEEFE. The 2003 budget was proposed in February. The National Academy of Sciences came out 10 days ago and identified Pluto in that regard. So, yes, you're right. It was not included because we didn't foresee that as being a highest-priority issue. Nevertheless even if it were, the question that we roll over in our minds is—it takes you 16 years to get there. Our bigger problem is, let us figure out a way to cut down that distance and time continuum problem with a power generation and propulsion issue or initiative. And a billion dollars is proposed in our program now to pursue those efforts that will be specifically designed to look at any of those destinations and cut down that time to inform the research agenda in a contemporary way for principle investigators within NASA, Science and Engineering Community folks, or elsewhere. And so our objective is to focus on that area.

SPACE STATION

The second area that you mentioned is on—as it pertains to Space Station itself. What we have just released, just gotten a very specific coverage of from a very thoughtful panel of experts of external scientists in every discipline that would have the opportunity to utilize the microgravity condition that is unique on International Space Station is to prioritize for us what are the opportunities for greatest breakthrough? Where are the areas we should concentrate our attention first and really put our time to it as we complete, as it is an engineering work in progress right now. We are halfway there. And we haven't even achieved the core configuration and had hoped to get to that within the next two years given the current systems integration challenges, and then to pursue the scientific agenda that our group of external scientists have provided in each of those major disciplines and how we prioritize and go about doing that first.

That is exciting. We have got a lot of principle investigators, a lot of scientists, who are now looking at that saying those are the kinds of things we want to go after, that is the area we ought to concentrate our attention. So we are trying to prioritize, be selective and focused about how we begin to utilize and to excite that next generation.

Ms. JACKSON LEE. My time is short, and I cannot engage in a debate with you, Mr. Administrator. But when you cut out a return vehicle, whether you have the loftiest of goals and aspirations on research, you certainly cut into any viability of a space program where you talk about safety issues. This is not a human capital issue as much as it is how you give incentives to people to even come to the program when they say that you are not really serious when you are cutting down a return vehicle, or in any event, you are cutting down on the ability to do research at the Space Station. And I just—before Mr. Roth——

Mr. O'KEEFE. May I comment on that point?

Ms. JACKSON LEE. I would be happy if the Chairman indulges me. I would just be happy to have you talk about——

Mr. WELDON. Mr. Chairman——

Chairman ROHRABACHER. Ms. Jackson Lee, you have already taken twice as much as your allotted time.

Ms. JACKSON LEE. Not any longer than anyone else. But I——

Chairman ROHRABACHER. Yes, you—Ms. Jackson Lee, you have taken three minutes longer than anybody else on the panel so far.

Ms. JACKSON LEE. I—can I get a number, Mr. Chairman, from him on the amount of cuts that he thinks he might have?

Mr. WELDON. Mr. Chairman——

NO NASA WORKFORCE CUTS ARE ENVISIONED

Chairman ROHRABACHER. Do you anticipate cuts at this time?

Mr. O'KEEFE. I don't see it as being an increase, don't see it as a decrease, don't have a target, none of those things in mind. Again, as we explored a couple of discussions here, this is an opportunity for the authorities and a flexibility to establish plans for your consideration as part of a budget submission that could contemplate increases or decreases. I thank you for your consideration before we proceed. So, I don't envision cuts.

Chairman ROHRABACHER. Thank you very much. Let me just suggest that if we ask large numbers of questions and five minutes is allotted to us, and it does take up time to have some answers to some of those questions, it might—it just might be better for us to have a dialogue back and forth asking one or two questions at a time because it does permit that back and forth that I think is beneficial for us to get to the heart of the matter. And I tried to be, as Chairman, pardon the expression, flexible, which seems to be the word of the day, in permitting people to have more time. And I know Mr. Gordon has a couple of questions.

Mr. GORDON. Mr. Chairman, you are going to allow a second round so we will have a chance to ask more questions?

Chairman ROHRABACHER. Certainly. That is correct.

Mr. GORDON. Okay.

Chairman ROHRABACHER. Now we have Mr. Lampson who is an activist as well, as you can tell by the panel so far that we have some people who are deeply active members of this Subcommittee and take a very personal interest in what is going on.

Mr. LAMPSON. True believers, Mr. Chairman.

Chairman ROHRABACHER. All right. And we have now one of the truest believers. Mr. Lampson, go ahead. You may proceed.

WHY JOIN NASA?

Mr. LAMPSON. Thank you, Mr. Chairman. Mr. O'Keefe, the Administration's budget plan for NASA will not keep up—it doesn't appear to be set to keep up with inflation over the next five years. The major research and development activities such as aeronautics are facing cuts this year. The future direction of many of NASA's programs such as Space Station and Space Shuttle are still to be determined. The Administration's budget requests—advocates privatizing more, if not all, of the Shuttle Program as well as contracting out other NASA activities.

Over the last 10 years as we have already heard, NASA's workforce has been cut from 25,000 to 18,000 employees. Given that outlook, why should a good scientist or engineer or project manager seek to join NASA or stay there if offered other options?

Mr. O'KEEFE. Our mission objective is to understand and protect the home planet, to explore the universe and search for life, and to inspire that next generation to explore. As we have arrayed our programs, to try to accomplish those three mission areas and stick to our knitting understanding those. We have talked about a few of those. And I think there is an excitement that I see and have heard from folks, again, in one example I used of the power generation and propulsion initiatives, looking at that as an opportunity that is unparalleled, that has not been pursued at NASA or elsewhere in a long time. And so as a consequence of the enthusiasm to want to join this institution and be part of this Agency's storied legacy is something that we are seeing a real resurgence in right now.

Mr. LAMPSON. Do the human capital legislative provisions that we are talking about address that, this basic problem?

Mr. O'KEEFE. It will help us get there. Yes, sir. We will believe that to be true.

Mr. LAMPSON. Mr. Roth, would you comment on this?

Mr. ROTH. Well, again, I—you know, I think we share the same goal. But I don't think this legislation furthers the accomplishment of the goal. A lot of it is pay. And there is no new funding streams. So, the expanded use of bonuses will come at the expense of other workers, current employees. The current use of scholars—the use of scholarships will come at the expense of current employees. We will not be able, I believe based on these proposals, to attract the best and the brightest. And I am afraid that you are going to end up losing the best and the brightest that you have. Flexibility is the key word of the day. And I want the Chairman to know I didn't want to be painted into a corner that I don't belong in.

AFGE PAY-BANDING EXAMPLE

We have spent time—I will give you an example, the Navy Sea Command. We spent a year with them on a pay-banding project. Mr. Weldon mentioned moving quickly through on pay. We had a terrific pay-banding project that we had with Navy Sea. And we could have sold it to the employees. It had categories of workers, administrative types, professional scientists. And when it came down to the bottom line, the people we were "negotiating with" agreed. But when we went up higher, we were told it has to be budget neutral. Well, that means some people in a major way are going to lose. We couldn't sell that. A pay-banding project where people are going to be getting paid more quicker cannot be budget neutral. In this scheme here, we do not see any new money. We don't see how we can attract people. And the voluntary incentive is to get the older or more mature workers out the door. They are specifically tied to the position being reduced. So, you are not then bringing in a new worker. So, again, I don't want to be painted into a corner. I think current demonstration project authority in five years, say 4703 with an agency of this size, would be appropriate for AFGE and NASA to explore some of these proposals. As I have said, we have talked about quite a bit of this with Senator Voinovich. So, we are not intransigent on the issues.

Mr. LAMPSON. Thank you.

Mr. ROTH. Sure.

Mr. O'KEEFE. Thank you for letting me—I think one very important distinction—and I apologize, Mr. Roth. This is an authorization request. It is legislative in nature. So, as a consequence we are asking for the authority. And again, I have to emphasize that the funding issue and exactly where the budget consequences will occur is—will be presented to Congress upon its use to be contemplated how it be used. And if you make the decision that that is an inappropriate use, that is the forum in which to make that determination.

HUMAN SPACE FLIGHT

Mr. LAMPSON. I am running out of time. And I want to get one more statement in here before I lose my five minutes. This has been my concern, as I was trying to explain through that. And I really believe that our Human Space Flight Program is adrift. I don't believe it has a clear vision or commitment to any goals after the completion of the International Space Station. And I have tried myself to put forth something of a vision by introducing the Space Exploration Act for 2002 which provides a concrete set of specific, incremental goals that are challenging and exciting. We need to reinstate a challenge similar to the one that we had during the Apollo Era. And I believe that that would enhance NASA's ability to recruit and retain that higher quality workforce much, much better than these kinds of things.

MOTIVATIONS FOR JOINING NASA

And I am struck by a couple of things here. I noticed that panel periodically. May I read both of these? "Where there is vision, people perish." I feel like the people that were coming to NASA during the time that we had our greatest accomplishments were not looking for a job. They were looking to dream a dream and build it. And I think that what we are doing today seems to be seeking to encourage those people to look for jobs and not dream dreams. And I think we are missing the boat.

Mr. WALKER. First, I think the primary reason that people come into government is not for money. It's to make a difference for the country. And in the case of NASA, I would say to make a difference for mankind. Secondly, I. . .

Mr. LAMPSON. Are we doing that? Do——

Mr. WALKER. Well, I understand.

Mr. LAMPSON. I feel like the things that I said a minute ago at the beginning, that our cuts that we have achieved, the changes, the downsizing of projects——

Mr. WALKER. A couple of comments. First, I think what you are commenting is, is there a specific objective, that that might help to rally the troops. Just as President Kennedy said, "We will put a man on the moon by the end of this decade," that could help potentially. But I think the idea of trying to make a difference for the country of mankind is something that will attract people. But you have got to be more specific. Secondly, I do think they need more flexibilities ideally government-wide. I think there needs to be adequate safeguards. But I do think they need more flexibilities. And thirdly, I could say that if you get the flexibilities, they have got to be funded. If they are not funded, it might not do much good. So, I think all those are legitimate points that ultimately Congress has to consider.

SETTING GOALS THAT INSPIRE YOUNG PEOPLE

Chairman ROHRABACHER. Thank you, Mr. Lampson. I agree with you, Mr. Lampson, that we are adrift in the sense that there isn't a goal. And I think that the new team at NASA has made it clear, however, that—and you will correct me if I'm wrong—that the NASA is not going to be driven by destination and that it is going to be more of a technology or a general theory rather than Destination Mars or Destination Moon. Now, I don't know. Maybe America and maybe the people need to have a destination to do know one thing. Mr. Lampson makes a very good point, and that is that—and it is not just this Administration. It is not a partisan issue at all, because I believe this is true of the last Administration and this Administration. And that is that the White House has neglected America's Space Program. And I believe in order to have an excitement and a feeling of a team and moving forward trying to achieve those objectives that we have, those objectives have to be outlined by the President of the United States.

Mr. O'KEEFE. True.

Chairman ROHRABACHER. And I don't believe we heard it in the last Administration. I don't believe we heard it in this Administration. And I would hope that Mr. O'Keefe and Mr. Walker who represent management skills of more than anybody I know. You were hired for management skills, of course. Dan Goldin came from a private sector and had great management credibility himself. But perhaps something as important as space and important as America's advancing into this new frontier and how it relates to the technological advances, perhaps it requires more than just management approaches. And we haven't had much of that. And I agree with Mr. Lampson on that. I may disagree with what destination he filled out. But I think we both agree that lofty goals and important goals are things that inspire young people. With enough said, let me mention we are going to go into a second round with permission of those who are with us today.

And first of all, let me note that I think that a trip to Pluto would take us how long, 16 years? So, under what we are talking about today, we might have—we might shoot of this rocket to Pluto and we would have a totally new NASA workforce by the time it got back. Nobody would even remember. "Oh yeah, we shot that off when old Harry used to work here." So, you know—all right. And there was some thoughts here—you know what I'm going to do, I'm going to let Mr. Gordon, as I'm organizing my own thoughts, proceed with some of the questions that he had and needed to follow through on.

SETTING PRIORITIES AT NASA

Mr. GORDON. I will try to be quick. Thank you, Mr. Chairman. Ms. Jackson Lee and Mr. O'Keefe, in your dialogue you brought up the question about authorities. And I think this is pretty much of a threshold question and more than we can get into today. But let me tell you what my concern is. And then maybe we can—I will suggest a potential remedy to help me better understand this. You were talking about establishing priorities in science. And I think that has pretty much been done. You have had a commission that has done that for you and ranked them. But most—but their highest priorities you are putting aside now, saying that they can be done better if we have more propulsion and a better propulsion. And so, we have got to put these aside. Maybe I am misunderstanding this. And that is why I want to better understand it. So, what I would like to do is submit to you some questions so that I can better understand this whole idea of prioritizing versus maybe—I guess you would say you're not undoing those priorities, you are just pushing them back so that you can do them better with additional propulsion. So—and I need to understand that. And I will submit some questions. Now going back to more of the topic at hand. While NASA's Strategic Resource Review was going on——

Mr. O'KEEFE. Congressman, I am sorry. Could I offer just a very brief comment?

Mr. GORDON. Sure. Absolutely.

Mr. O'KEEFE. What you just defined as a goal, an enabling technology goal to go anywhere, and if we can figure out how to do that in a span of time to be early enough to inform the current generation of research, science, and principal investigators, that is an accomplishment. Then we can talk about where those destinations are. But right now if we have to sit back and wait for 16 years for the results. . .

Mr. GORDON. Well, there is some controversy as to whether you really save that much time in terms of the starting and stopping. But also, we get to the point if you determine, okay, if we have these priorities and you say, well, let's wait on these priorities. We won't do anything now until we get better propulsion. Well then when we get better propulsion that takes us—let us just assume that you will go from 16 to 12 years—then why not just say, well, you know, if we were to work a little bit more on this, maybe we could take it from 6 to—from 12 to 6, and let us do a second theory. So, we could be sitting here for those 16 years continuing to postpone, because we want to get better. Now, you know, everything is a balance. And somewhere there is, you know, you wait two more years if you can get 100 percent increase, you know, you don't wait two more years to get a two percent increase. Now, you know, somewhere in between those decisions have got to be made. I would just like to have a better understanding of your thinking in those terms so that we can make those determinations or do my job anyway.

NASA'S STRATEGIC RESOURCE REVIEW

Now, back again to the issue at hand. While NASA's Strategic Resource Review was going on, Congress was told that it would be "an exhaustive review, assessing the future roles and the missions of each NASA Center, what facilities should be closed, what NASA activities would be contracted out, and what the future workforce levels should be." We were told that "the SRR recommendations would drive the restructuring of NASA and its workforce." NASA was going to provide the SSR recommendations to Congress with the fiscal year 2003 budget request in February. It is now July. Since you have now sent over NASA's workforce restructuring legislation, I assume you have decided which of the SRR recommendations you will adopt. So, when will you share the SRR report and its recommendations with this committee?

Mr. O'KEEFE. Matter of fact, we have just met on this in the last few days. I regret to inform on this one, Congressman, that the expectations that were set in motion, I don't see evidence of it. The SRR is——

Mr. GORDON. I am sorry. You said you don't expect what?

Mr. O'KEEFE. I don't see evidence of the lofty expectations you enunciated at the beginning in terms of what was intended when this was launched off almost two years ago, I guess, with the intent of trying to work through a series of things to inform the kinds of decisions you just talked about. I don't see evidence of it. So having spent the last six months looking at what are the elements of each of the Strategic Resource Review aspects, it is not going to be as comprehensive as you defined. What we will be endeavoring to do here——

Mr. GORDON. Now, why is that? Is it that there is not the ability to accomplish that?

Mr. O'KEEFE. No. I just—I don't believe that the objectives that were set and, in turn, how it was implemented to look at what infrastructure would be facility, capacity, all that was actually accomplished. So, what I have reviewed in looking at——

Mr. GORDON. So, were those not good ideas in the first place?

Mr. O'KEEFE. They may have been. But they certainly aren't in evidence right now.

Mr. GORDON. And when you say not in evidence, are you saying not in evidence as to the fact that that information is being brought to you or not in evidence as it needs to be done?

Mr. O'KEEFE. Not in evidence as being brought to me.

Mr. GORDON. Okay.

Mr. O'KEEFE. What I see——

Mr. GORDON. Well now, you are in charge now.

Mr. O'KEEFE. Yes, sir.

Mr. GORDON. So, do you—I guess, do you expect to put the people to work to bring that evidence to you?

Mr. O'KEEFE. Yes, sir, on specific selected areas of it because the very comprehensive nature of what was intended by this, the SRR, has not been accomplished.

Mr. GORDON. So, is it not needed or—is it needed or not needed?

Mr. O'KEEFE. I think we need to implement the elements of the SRR that I think are coming to fruition here that are really modest by comparison to that. And they certainly aren't comprehensive enough to——

Mr. GORDON. Well, do you have the managerial ability to get it done?

Mr. O'KEEFE. I hope so, sir. I am certainly trying.

Mr. GORDON. So, then are you going to do—so, are we going to get an SRR? I mean, you said that you are not getting it because people aren't providing it for you——

Mr. O'KEEFE. I am sorry.

Mr. GORDON [continuing]. Yet you are in charge now.

Mr. O'KEEFE. Yes, sir.

Mr. GORDON. Are you going to put the mechanism together for us to get a good one?

Mr. O'KEEFE. Yes, sir. We will release a report here in very short order. I am expecting no more than four to six weeks max. And I think what we will see in that report is a modest set of proposals and efforts that are very specific to very specific facilities. And it is not as comprehensive as what you——

Mr. GORDON. Okay. So, in four to six weeks we are going to get the modest version?

Mr. O'KEEFE. Yes, sir.

Mr. GORDON. Now—but what I am asking you is——

Mr. O'KEEFE. We are going to close it down.

Mr. GORDON [continuing]. And I think that what—you are going to close it down?

Mr. O'KEEFE. Yes, sir. I think we are going to terminate that effort and move on.

Mr. GORDON. But I thought you said that it was important information for us to have.

Mr. O'KEEFE. Yes, sir. I am going to give to you what the information I have received.

Mr. GORDON. Right. I know. But you are going to give us the modest one. But I—excuse me. Let me just—I want to get this straight.

Mr. O'KEEFE. Yes, sir.

Mr. GORDON. Was the original objective important?

Mr. O'KEEFE. Again, having not been there when the original objective was stated, how they determined to get there, I can only go through that as a revisionist historian.

Mr. GORDON. Okay. But from what you—would that be good information for you to have?

Mr. O'KEEFE. Yes, sir. And I have gone through it. And again, the best I can do is be a revisionist historian on what was the——

Mr. GORDON. Okay. Well, I am not trying to go back. You are in charge now. We are not trying to go back to what they meant then. You are in charge. I am trying to determine whether or not this is important information and whether or not you intend to put it together.

Mr. O'KEEFE. Yes, sir.

Mr. GORDON. You do. Okay. Now, again, I think I don't want to be talking—sorry——

Mr. O'KEEFE. Yes, sir.

Mr. GORDON. What you are saying you are going to put together is you are going to give us this modest version. Now, is it important that you use your ability then to get the more elaborate version for us? Is that important information for us to have?

Mr. O'KEEFE. I am terribly sorry to have been confusing, sir.

Mr. GORDON. Okay.

Mr. O'KEEFE. What will come forward is not as comprehensive as what you define in your——

Mr. GORDON. I understand that.

Mr. O'KEEFE [continuing]. In your——

Mr. GORDON. I understand that.

Mr. O'KEEFE [continuing]. In your pre-ambulatory statement in asking the question.

Mr. GORDON. All right.

Mr. O'KEEFE. So, it will not have the——

Mr. GORDON. I understand that. You said that. So, I am not doing a good job. Let me try again because I understand what you are telling me. But I am not being clear with what I am trying to communicate to you. I understand that what you are saying is that you are not going to give us as comprehensive a review as was stated in what we see or what our view of the original goal. I understand that. Now what I am saying is, you are in charge. And if that original goal and if that comprehensive goal that I set forth, if that is good, I would like for you to tell me if it is good. And if it is good and important, do you intend—not out of information you have if you don't think it is enough there—but do you intend to go forward to try to produce that information on your own?

Mr. O'KEEFE. Yes, sir. We will give you what we have. And I will——

Mr. GORDON. Okay. Well, you go ahead and say that. And then I will have to say it back. But you go ahead and say it. Say what you said again. Go ahead.

Mr. O'KEEFE. We will give you what has been the result of that specific Strategic Resources Review in short order.

Mr. GORDON. Okay.

Mr. O'KEEFE. We will then continue with our series of things that we have discussed several times in public hearings, as well as privately, on a matter of series of business issues that pertain to the business case for Shuttle competitive sourcing, what the utilization and maximization for the International Space Station will import, whether or not there will be a non-governmental organization that will be asked to sort those scientific priorities. We will do these on a business-case basis and look at that. And that is what we discussed——

Mr. GORDON. And when would we expect to get those?

Mr. O'KEEFE. I think over spans of time.

Mr. GORDON. And what span of time would that be?

Mr. O'KEEFE. Certainly to accompany the fiscal year 2004 budget submission you will receive in February——

Mr. GORDON. Okay. Then let me go back to——

Mr. O'KEEFE [continuing]. No later than that. All the elements will be coming in from Italy. But——

Mr. GORDON. Okay. Well, let me just go back because we have got a short period of time here.

Mr. O'KEEFE. Yes, sir.

Mr. GORDON. Do you think it is important to have a review that assesses the future roles and missions of each NASA Center, what facilities should be closed, what NASA activities should be contracted out, and what future workforce levels should be? Do you think that would be important information for you to have?

Mr. O'KEEFE. Yes, sir. And the approach we are taking to that is the Strategic Management of Human Capital Plan that we have put forward and have developed that is now part of what is guiding us to a series of proposals here as well as the other three initiatives I talked about.

Mr. GORDON. Okay. That—okay. So if I were to look at that plan, and which I think is more—you know, and if I was going to look at that plan, and I wanted to know the mission of each NASA Center, the future roles and mission of each NASA Center, and what facilities would be closed, and what NASA activities would be contracted out, would I be able to find that in that information that you are sending me?

Mr. O'KEEFE. No, sir.

Mr. GORDON. You wouldn't. Okay. Now, is that information that we should have?

Mr. O'KEEFE. It is not something we have developed. It is not something I have commissioned.

Mr. GORDON. Right. I understand that. But what I am trying to say is, one, is it important, and I would just like a yes or no, it is—you know, is it important or not, and two, if it is, do you intend to try to gather that information?

Mr. O'KEEFE. Yes. What is important is understanding what the infrastructure currently is and how it will be utilized. And that is what we present to you in our annual budget, sir.

Mr. GORDON. Okay. Well, we can—you know, we can keep doing this back and forth, I guess, for some time, you know, because at first you could pretend like you didn't understand me. But I am trying to be real clear here.

Mr. O'KEEFE. Yes, sir.

Mr. GORDON. And so, I will try—I will say once again——

Mr. O'KEEFE. Please let me assure, you sir, I am not pretending.

Mr. GORDON. Okay. So, you are not pretending you don't understand me then?

Mr. O'KEEFE. Yes, sir. I am not pretending.

Chairman ROHRABACHER. For the fairer part, I don't understand you either.

Mr. GORDON. Okay. All right. Well, again—well, let me try to say it again because, again, that is what we are trying to get worked out here. As I understand, the original NASA Strategic Resource Review was going—when it was going on, Congress was told that it would be an exhaustive review assessing the future roles and missions of each NASA Center, what facilities would be closed, what NASA activities would be contracted out, and what future workforce levels should be. Now, I understand that the information that has percolated up to you will not be able to answer those statements that I just read and you are going to give us the best that you have. I understand that. So now my question is, is it—would it be good for Congress and NASA to have a review that assesses the future roles and missions of each NASA Center, what facilities should be closed, and what NASA activities could be contracted out, and what future workforce levels should be?

Mr. O'KEEFE. I am not currently planning that, sir.

Mr. GORDON. Okay. I understand you are not planning it. Now, my question—the first question was, would that be good and important? Would that be helpful to us?

Mr. O'KEEFE. It might be. But I think it is asking the—slightly the wrong question. The first question that has to be asked before you get to that level of indenture is what are the missions of the Agency overall and how would you pursue them as collaboratively as you can between and among all the centers as well as the activities that we are engaged in, and then look at what does it take to carry that out in terms of resources, capabilities——

Mr. GORDON. But when would you expect—you say the first part of that will be done by the '04 budget request, by the——

Mr. O'KEEFE. The objective is to present to you as part of each of our annual budget requests something that more comports with that philosophy.

Mr. GORDON. Okay. Well, then once you have that, then you will be able to go to the next step, won't you, because didn't you say that you couldn't go to the step that I just requested until you had the previous step, but now you are going to give us that previous step with this next budget?

Mr. O'KEEFE. Again, I am not trying to be evasive, sir. I just——

Mr. GORDON. Okay. Well, then let us just keep—you know, and I am not trying to be argumentative or unclear. And we will just, you know——

Mr. O'KEEFE. I appreciate that.

Mr. GORDON. I don't mean to take too much time. But I don't feel like it is appropriate to be rope-a-doped, you know. So, we will take the time. . .

Chairman ROHRABACHER. The Chair is trying to give you your time.

Mr. GORDON. Yeah. We will take the time and try to get it figured out.

Mr. O'KEEFE. Mr. Chairman and Congressman, I am not trying to do that at all.

Mr. GORDON. Okay. I know. I don't think you are either. That is why we are going to take the time and try to get this figured out.

Mr. O'KEEFE. Yes, sir. Again, it is a matter of approach and how you do this.

Mr. GORDON. Exactly. Fine.

Mr. O'KEEFE. And the one I have suggested to you is that the method we are attempting to use——

Mr. GORDON. Right, right, right. And you are going to get that done by this fiscal year—I mean, the next fiscal year summary. That is——

Mr. O'KEEFE. I am sorry. The difference is the approach that we are talking about is preparing the various objectives and approaches programs that then are manifest in the budget before you. And when the '04 budget comes forward, the '05 budget comes forward, it gives you more visibility on what that overall planning objective will be. Then from there, it informs a bit more and tells us a lot more about what infrastructure capacity is required.

Mr. GORDON. So, we are not going to know that until after the fiscal year 2005 budget?

Mr. O'KEEFE. No. Again, as each budget successively comes forward, our attempt is to give you the strategic plan for what we would like to——

Mr. GORDON. Okay. Well, great. Good. Okay. So, after—so, you are going to have the first part of that. It should be coming up here in just a few a months with—and you are going to give us your strategic plan after you get the fiscal year 2004 budget proposal. Is that correct?

Mr. O'KEEFE. Yes, sir. Then update that——

Mr. GORDON. Okay. And then what would you expect for that strategic—how far would you expect for that strategic plan to go?

Mr. O'KEEFE. As it typically goes, it is five years.

Mr. GORDON. And what would you expect to tell us in that?

Mr. O'KEEFE. We are formulating it as this juncture right now. But I think we have talked a little bit about a few of them here. The first one I can tell you with certainty is we are really going to look at enabling technologies and try to determine how——

Mr. GORDON. Okay. So—and I will close right here. So, when can we expect that?

Mr. O'KEEFE. As part of the '04 submission for sure. And as a series of things between now and then that we have talked about several times——

Mr. GORDON. And we will get that strategic review after that? I mean, you were saying you need that information to make your strategic review.

Mr. WELDON. Mr. Chairman——

Chairman ROHRABACHER. Yes?

Mr. GORDON. I mean, I just want to know——

Mr. WELDON [continuing]. This is a very important conversation. But I have a press interview. And I——

Mr. GORDON. Well, you go right ahead. I—let me yield. Okay. I will, you know, yield for a while. You go right ahead.

Chairman ROHRABACHER. Let me take the Chairman's prerogative. And we will be happy to give Mr. Gordon a couple more minutes at the end after some other——

Mr. GORDON. I want to say that it is just, again, we are on Mr. O'Keefe's watch. I understand that maybe this past one, the time slipped. It is not doing good. But now you are in charge. Management is your watchword, you know. And that is why we would like some timetables and to know what is going on and things that we can depend upon.

Chairman ROHRABACHER. Thank you very much, Mr. Gordon. And other members do have other things going on. That is why we have to bring in the conversation. Go ahead, Mr. Weldon.

NASA'S BUYOUT AUTHORITY PROPOSAL

Mr. WELDON. I thank you, Mr. Chairman. Mr. O'Keefe, NASA's proposal, I believe, asks for permanent buyout authority. Is that correct?

Mr. O'KEEFE. Yes, sir.

Mr. WELDON. In the 104th Congress, the Civil Service Subcommittee under then Chairman John Mica held hearings. Witnesses included experts from the private sector who described how buyouts could be used most effectively. In that testimony, they recommended that buyouts be offered as a one-time opportunity, that plans should be developed in secret, and that the employees should have a short time to accept the offer and leave. Doesn't your proposal contravene some of those recommendations? And isn't it possible that employees, particularly very senior employees perhaps who have stayed beyond their eligible retirement age, would continue to linger in a position in the hope of continuing to work longer and then ultimately take a buyout as they go out the door?.

Mr. O'KEEFE. No. The intention here is to—and I understand the point. That is a working, you know, philosophy that is one of the challenges of dealing with buyout authorities. And the way it has been implemented best, as I gather it from both Mr. Walker as well as through the Office Personnel Management, is to target the buyouts to those disciplines or workforce excesses where you see that the workload that is required there isn't supporting that activity and then utilize that capability in areas where you are underutilizing or have a shortage of capacity in those cases and utilize it for the purpose of recruiting and retaining. So as a result, our approach would not be to do this under any circumstances in secret. The approach would be to be very clear about how the targeted buyout authorities would be implemented using the best practices of the pilot programs that have been out there for years.

Mr. WELDON. You have to understand that the testimony that we received ran contrary to that though, contrary to the way you are proposing it.

Mr. O'KEEFE. Yes, sir. I take it from your comment there. And that is——

Mr. WELDON. You know, as we move through all this, I would like to sit down and talk with you perhaps off-line a little bit more on that. Did you want to add anything to that, Mr. Walker?

GAO EXPERIENCE WITH BUYOUT AUTHORITY

Mr. WALKER. I would. In 2000, the Congress gave the GAO a term buyout authority to realign the workforce for a period of years and term early-out authority. My personal view is you should treat those differently. I think the idea of having long-term flexibility to offer early-outs to realign the workforce makes sense. Buyouts, I have great concerns about because if you don't handle it the right way, you end up freezing the workforce in place waiting for the next buyout offer, if you will. We have not used the buyout authority at the GAO. We have used early-out authority. I think if you are going to provide buyout authority, you may want to think about having it for a period of time and be very clear in communicating to the workforce what management does and does not plan to do with regard to that authority to achieve the desired outcomes.

NASA'S WORKFORCE BONUS PROPOSAL

Mr. WELDON. Thank you. Mr. O'Keefe, your proposal calls for bonuses higher than those currently allowed under current law because the bonuses, in effect, now are described as ineffective. As I understand it for bonuses to work effectively and specifically to affect performance, they have to be well designed and there has to be a credible system for measuring performance and managers who are willing to use that system to make meaningful distinctions between workers. Has NASA taken any steps to develop those types of performance systems for managers to use?

Mr. O'KEEFE. Yes, sir. We are beginning to work through that as part of the strategic, you know, plan that we have developed for the human capital objectives to define what those specific performance criteria are and to base the buyout approach that we are looking at as well as the recruitment and retention objectives on a performance basis as well.

Mr. WELDON. Mr. Roth, did you want to comment on that at all?

Mr. ROTH. Well, I think you hit it pretty much on the head. There is two problems. One is the money. It is not separately funded. So, you are not getting people using the bonus authority that already exists. So, to say it is insufficient really is a little unfair. It is not funded. And so, people are not using the current authority. The other problem that I am—I think you hit exactly on the head is that it is not just employees. It is also managers don't believe the current systems have credibility. So, they are not willing to distinguish employees and say this person is entitled to a bonus, this one is not, because the systems don't have credibility. The unions are left out of negotiating or developing jointly in most agencies' performance management system. So, we can't sell them to the employees. And they seem very subjective to employees.

Mr. WELDON. Well, let me ask you an important question though. If the Agency were to develop proper management tools, and if the Congress were willing to fund, and the Administration were willing to put the proper amounts of money, would the union support the use of this kind of tool in workforce shaping and management?

Mr. ROTH. I am not going to address the amount that they are asking for because I think it is quite high. But the idea——

Mr. WELDON. The principle.

Mr. ROTH. I don't want to talk about the amount.

Mr. WELDON. Yes.

Mr. ROTH. The principle, with the caveat, just like in the private sector, you must have union involvement to sell it to the employees. It is very controversial. Where we have done it off-line, in collaboration, or as a permissive subjective bargaining, or in a demo project, we have had success. Where we have been delivered a complete package unilaterally, management cannot do it unilaterally. They can not sell it to the workforce. So, my answer to you is, yes, if there is strong union involvement in creating the system. And we have done it before.

Mr. WELDON. Is that one of the issues that you have been working with Senator Voinovich on in your negotiations with him?

Mr. ROTH. I don't believe that—he has taken a lot out of his proposal. I am not quite sure. Do you know? Yeah. We are working with him on that. But we also worked with him to get out this idea of the one-for one, if someone leaves, you can't replace him.

Mr. WELDON. I would be very happy to work with you and the Administrator. This issue will have to come before the Committee that I chair. And if we can come to an agreement, I think this is an area where it could be very useful for agencies like NASA to not only recruit more talented people, but more importantly, keep them. I know my time has expired. But the Chairman is using a very loose gavel today.

NASA'S DIRECT HIRE PROPOSAL

And let me just ask one more follow-up question. And then I do have to run. I share your concern about direct hiring and that bringing—Mr. O'Keefe, bringing people in can sometimes be very difficult. I have been informed through the Subcommittee that the U.S. Geological Survey has developed some technologies that allow them to bring people in more quickly. I don't know what those are. But have you looked at those and is NASA employing any of those methods right now?

Mr. O'KEEFE. Yes, sir. We are trying desperately to reduce the time that it takes to do that. And USGS is one of the leaders out there in trying to figure out how to work that through. And plus, I have got to say that Kay James has done a remarkable job at OPM of trying to really address that particular set of questions. They are, nonetheless, a series of specific limitations that will always be inherent and will take time, that require a vetting process, then a selection of top three, and a whole range of different approaches to this. But frankly, in the approaches that Congress has enacted previously of permitting direct-hire authority, they have been able to reduce that time down to 15 and keep the quality very, very high without suspending Veteran's Preference, without really looking at the kind of important policy objectives we have always felt were significant. So, it can be done. And there are great benchmarks on how to go about emulating their best practices in those areas as well.

Mr. WELDON. Did either of the other witnesses want to elaborate on that a little bit before I go?

Mr. WALKER. Just very quickly. You have put your finger on a very important point. If we are going to move to a system that recognizes, rewards, and promotes people based on skills, knowledge, and performance rather than the passage of time and rate of inflation, which is really what I think the Administrator is talking about, we have to have modern, effective, credible performance management systems. The current performance management systems at NASA and most of the government do not meet that definition. That does not take a change in law. Right now, leaders and agencies can end up, you know, working on to achieve that objective. They should do that. And they should work with labor to the extent that they have an organized workforce. And those systems should be validated, which there is a process to do it, in order to minimize the possibility of litigation. We have done this at GAO. We are more than happy to share what we have done with you and with others, Mr. Weldon. Thank you.

Mr. WELDON. Thank you.

Mr. ROTH. I agree with what the Comptroller General said. I just want to also point out there are a lot of self-inflicted wounds in the hiring place, this so-called vetting process. This is all internal to the Agency, having to do the position description, having to get the approvals, going up to OMB for the authority to fill a slot. A lot of this time is self-inflicted. It could be done away with.

Mr. WELDON. It is not a union issue.

Mr. ROTH. It is not a union issue. And it could be done by—it could be done even with keeping the rule of three, which we think is important if you want the best and brightest and they have the rosters. You should be able to go immediately to them and pull them right off them.

Chairman ROHRABACHER. Sounds pretty flexible to me.

Mr. WELDON. Mr. Chairman, I just want to thank you for calling this hearing.

Chairman ROHRABACHER. All right.

Mr. WELDON. And I think all of the witnesses have been very, very informative.

Chairman ROHRABACHER. That's a yes. And now, Ms. Jackson Lee, you may proceed.

ENTHUSIASM FOR NASA

Ms. JACKSON LEE. Mr. Chairman, I have had an affection and appreciation for NASA preceding my coming to the United States Congress. And I still maintain that enthusiasm. And so, this hearing greatly disturbs me because I think it is infused with a question of numbers rather than vision and mission. I am delighted that there are programs that are reaching out to ensure that we have a better reach to young people who may be interested in this mission. But I indicated in my previous series of questions that the mission is now fuzzy. And I am still not that much more clear on where the Administration deems itself going. And might I say, Administrator O'Keefe, my criticism is balanced inasmuch as I did not like the direction of numbers that the preceding Administrator was headed in, though I do acknowledge the great work that he did, as I will acknowledge the great work that you are doing and probably will do in the future. I am reminded of your visit to Houston when you expressed the awe that you had as a young person for the Space Program and, I assume, the continuing awe that you now have. Might I simply say to you to clean up the record, the Senate put money into the Pluto Program over the objections of the Agency and Administration last year because it had been known that this was a high-priority project. And might I say that one of the other sites was the Kuiper Belt, which was also part of the establishment of priorities. And it seems as if we are not going there as well.

NASA'S ROLE IN HOMELAND SECURITY

Let me, however, pose my questions—series of questions to Mr. Roth as I did not give you the opportunity to finish. The theme running throughout this is the issue of flexibility. In a homeland security hearing in this committee, I had such a belief in NASA's purpose that I had an amendment that said that the homeland security secretary should consult with NASA for its utilization in homeland security issues. I believe NASA has a great role and the research that it has done, and its satellites, and the International Space Station in homeland security are, in essence, informing us and providing us with the opportunity to be safer. That brings to question the point that you made earlier about knowing who all of our government employees are, but possibly not knowing who all our contract employees are with, again, the qualifier that I believe the contract employees have been a very useful resource for us. I do believe that is an important security issue.

And I would like you to comment on that as well as the fact that I would like you to comment on the Section 203 that I brought to your attention, Mr. Roth, before that deals with the Industry Exchange Program, if you are taking notes, and Section 204 that deals with the issue of authorizing direct hire for positions and critical needs in severe shortage categories. And I would leave it up to NASA and OPM to define those categories.

WHISTLEBLOWER PROTECTIONS

And let me just say one other thing that truly concerns me. I do not know whether flexibility means that whistleblower protection is lost. I do not know whether flexibility means that the employees that have been employed by NASA who are government employees whom I hope we don't label as employees who are ne'er-do-wells. I hope it does not mean that those employees who have worked to get up the ladder, some of whom are minorities, who have not gotten up the ladder, who are not in management evident by looking at the management levels in our centers, I hope that doesn't mean that we discard them. I hope it does not mean that the 1987—I believe that was the year. Let me just take the year away—it seems that that was the year of the Challenger incident that impacted so many of us, this nation, but those of us who live in Houston in the most devastating of ways. Our neighbors were lost. And one of the findings was that there was no communication, that there was no translation of the problems between contract employees and those in the government. No resources came to correct the data that determined that there was a problem with the O-ring, there was a problem with the booster circumstances or equipment, if I will. And so, the question is whether flexibility adds to the mission, adds to the safety, adds to the commitment, adds to the improving of making this workforce the best that we can get in the mission of NASA, or adds to saving lives? Do we have employees that will stand up, that will tell the truth, and stand by what they say with the proposals that we have before us?

Mr. ROTH. Well, we feel very strongly that the answer to that is no. The Comptroller General said—I don't want to misquote him—that there is a need for flexibility, but you must have strong safeguards. So far what flexibility has meant for the Transportation Security Agency is no union rights and limited, unenforceable whistleblower rights at the voluntary agreement of the Deputy Secretary of Transportation, not full whistleblower rights. We don't know yet about EEO rights. What flexibility has meant in the House in the Homeland Security Act is no collective bargaining rights. So when I hear leave the flexibility, let us forget about the details, leave it up to the administrator, leave it up to the director, leave it up to the secretary, I am hearing no whistleblower rights because this is the history so far—recent history, no union rights, no EEO rights. And of course he will say today, no, that won't happen. But when we ask for a guarantee that this will be maintained we are told, no, we need the flexibility. That probably or may not include union rights, EEO rights, or whistleblower rights. I just want to say that whistleblower rights and union rights go hand in glove. No sane whistleblower is going to stand up under current law on his or her own and say I saw something terribly wrong in this program. They would be fired. And this is through history proven. Without an organization like the union to run the interference like they did on the Border Patrol several months ago where two of our local presidents stood up and said we have a serious problem on our border and it is not being taken care of, they were proposed for removal. The union had to get them back. So when I hear flexibility, without the safeguards being guaranteed in the statute, I have serious concerns.

IPA PROGRAM

The IPA Program, there may be some value to it bringing in some private sector people. But I think the Administrator here—is here today somewhat to sell the Administration's proposal on Freedom To Manage—and again, I—without offending anyone I will say Freedom to Manage Act—because he makes it in his proposal government-wide. Why would it be government-wide if it is a NASA program? Do you want people coming into private industry into the most secret conclaves of our government and then going back out with this information? And then you don't even have a contractor inventory, knowing who your contractors are, who they have employed, and the numbers, types, and grades of these contractors. I just don't think that it helps NASA or any other government agency in maintaining the best and the brightest. And frankly, I have been a little disappointed that no one has talked about the current workforce being a pretty darn good workforce. I was going to say a different word with a "d". But an excellent workforce currently.

DIRECT HIRE AUTHORITY

Ms. JACKSON LEE. I have. I posed—I made the statement early on. And I would like you to respond to that, to answer my question about moving up the ranks. And these who have been there who have worked, what does this new flexibility do to their, if you will, morale when they hear this in terms of what their opportunities will be?

Mr. ROTH. Well, I imagine when they get leap-frogged they are going to be pretty upset about it. On the direct hiring, again, I am not sure—I haven't seen enough evidence that it is essential to NASA and to make it a better agency. People keep wanting to avoid the issue of pay. There is—you know, the goal which you have talked about, having a goal and a mission and a President that makes people want to serve and not having anti-government politicians bashing every Federal employee as a bureaucrat. That is important. And I grew up in the 60's. And government service was an honor and viewed that way. We were not used as whipping posts every election. But now, pay is a serious problem. And it has to be addressed across the board, NASA and otherwise. You can't avoid it. They are not going to get where they want to go with these scattered-shot bonuses, unilateral bonuses, selective bonuses.

They have got to pay their current workforce and recruit the new workforce. And it has got to be more pay. And I am not sure direct—that you need direct hiring if you get rid of all of the other internal agency problems on hiring someone; the supervisor being afraid to go, then the next highest level, and then finally outside the Agency to OMB where it may languish. They have these registers. They are available. If you cut out the internal haggling, I don't think you need the direct-hire authority.

Ms. JACKSON LEE. Mr. Chairman, I would like to ask Mr. O'Keefe. And I would just like to reframe the question more specifically.

Chairman ROHRABACHER. Ms. Jackson Lee, you have already had twice as much time again than we generally allot. And—but if we could have Mr. O'Keefe and Mr. Walker just comment on your last question and then we will move on.

Ms. JACKSON LEE. I would appreciate. Let me just clarify the point——

Chairman ROHRABACHER. All right.

WHISTLEBLOWER PROTECTION

Ms. JACKSON LEE [continuing]. Mr. O'Keefe, with respect to the question of the Challenger. This is not an attempt to reinvestigate. But it is an attempt to make, if you will, pertinent the inquiry about flexibility. And the real question is, who knows who was there amongst those employees that could have given information as to who did not act or respond to information they received? Does flexibility mean that perspectively we have that question always reigning above us inasmuch as flexibility means no whistleblower protection or flexibility means no incentives to stand up and speak up for what you believe in? What does flexibility mean?

Mr. O'KEEFE. Well, thank you, Congresswoman. I think you are hitting a very, very important point. And let me please reassure you there is nothing, nothing in our legislation to reduce current union, whistleblower, or equal opportunity processes, or safeguards, no intention whatsoever to change those at all. And so as a consequence, I think the flexibility in my mind means, and the intent in which we intend to use it, if I could borrow a phrase that the Comptroller General used in his opening statement, that we have an Industrial Age driven set of policies that exist today on human resource science and engineering and technology-related fields that are more dominant than almost any other agency in the entire Federal Government. It is an extraordinarily qualified, very competent, very skilled workforce that is in danger of being not of that condition five years from now, given the age of the workforce as well as the cohort that follows behind that are not as interested in pursuing it.

On the safety question, let me simply reassure you that the Rogers' Commission taught us a lot. One of the things I found most impressive from the first moment I walked into the Agency was that the events after January 1986 and the tragedies of Challenger is something that folks can recite in our agency absolutely chapter and verse in terms of what emerged from the Rogers Commission to assure accountability. And that is, accountability among all who are associated with the activity. We are all public servants in this regard. Whether they are direct hire, GS and public servants, or whether they are part of a contract effort specifically to engage in the assurance of safety of flight.

Our efforts since that time have been at least parallel to that which we have seen in the unblemished record of the Navy nuclear activities that have gone on since that period and before, and we view that as an admirable record and one that we continually benchmark against them. The question of accountability is a very clear one. It is the strongest legacy of the Rogers Commission, which we live with to this moment.

Chairman ROHRABACHER. Mr. Walker, do you have your——

Mr. WALKER. Yes. Three very quick points. First, we need to invest in our current workforce, but we need to prepare for the future. Secondly, we do not need to know who the contractors are by name for purposes of cost and quality. We need to know results. We do, however, need to know who they are for purposes of security. And last but not least, you mentioned about change. I agree that not all change is good. However, in today's world, without change, extinction will follow.

Chairman ROHRABACHER. Thank you very much.

Ms. JACKSON LEE. Gentleman, I thank you for your indulgence. I just simply want you to know that there are employees at NASA who are very, very fearful of the proposals and very, very unsure whether their talents are going to be utilized.

Chairman ROHRABACHER. Ms. Jackson Lee, you have had more time, approximately twice the time the Chairman has had, to express these ideas that you have, and I respect your right to do so. But in the future, you are just going to have to be more disciplined in being able to ask a specific question, not run over like this. I'm sorry.

Ms. JACKSON LEE. Chairman, it is only Congresswoman Jackson Lee that gets cited by the Chair, and she will continue to be cited——

Chairman ROHRABACHER. That is because only Ms. Jackson Lee is being discourteous.

Ms. JACKSON LEE. And she will continue to be cited and as well——

Chairman ROHRABACHER. All right.

Ms. JACKSON LEE [continuing]. Continue to work on behalf of her constituents.

Chairman ROHRABACHER. Ms. Jackson Lee has been discourteous and way——

Ms. JACKSON LEE. I think the comment is unnecessary.

Chairman ROHRABACHER. Well, the Chair is taking over right now. The bottom line is we need to be courteous to each other.

Ms. JACKSON LEE. And that I am.

The. CHAIRMAN [continuing]. And it is——

Ms. JACKSON LEE. I am the only——

Chairman ROHRABACHER. Ms. Jackson, you are not recognized.

Ms. JACKSON LEE. For you to cite continuously——

Chairman ROHRABACHER. Ms. Jackson Lee, you are not——

Ms. JACKSON LEE. And I am not going to accept it. Thank you, Mr. Chairman.

Chairman ROHRABACHER. Ms. Jackson Lee, you are not recognized.

Ms. JACKSON LEE. I am the only Member that you comment on. It may be that I am the only African American woman sitting here.

Chairman ROHRABACHER. Ms. Jackson Lee, that type of charge is beneath you, and is beneath your dignity to make that charge, Ms. Jackson Lee. Ms. Jackson Lee——

Ms. JACKSON LEE. Thank you for your courtesy.

Chairman ROHRABACHER. Ms. Jackson Lee, I wish we could say the same——

Mr. GORDON. Mr. Chairman, we ask——

Chairman ROHRABACHER. Yes, sir.

Mr. GORDON [continuing]. There be an opportunity for all of us to continue to ask questions——

Chairman ROHRABACHER. That is correct.

Mr. GORDON [continuing]. If we go into a rotation.

Chairman ROHRABACHER. I would note that the Chairman has been enormously generous with the time on this particular panel, and that is something that perhaps the Chairman will have to reconsider in the hearings in the future. Mr. Lampson, I did not take my second round of questions, but I will let you proceed, and I will end up the hearing.

RECRUITING THE YOUNGER GENERATION

Mr. LAMPSON. Thank you, Mr. Chairman. I will very likely give you back some. I wanted to ask Mr. O'Keefe if you had the opportunity—right at the very beginning, you made the comment about the young folks that are here today from universities. Have you talked with them or others and asked them some of the questions that you are presenting to us today? What they envision for the future and what would attract them there?

Mr. O'KEEFE. Yes, sir. No. I have not had the opportunity to meet with this group, but I have got an opportunity right after this hearing to do so, and they have been very patient waiting for that. So I am hopeful they will be indulgent. But I have had the opportunity to talk to groups that—university students and particularly to grade school kids, too because it is—I have come to realize, I guess, that it has been impressed upon me by educators is if you can motivate interest in math, science, engineering, technology-related kind of activities in the grade school, period, their likelihood of pursuing those kinds of fields in college and sticking with it is much, much higher.

Because if you think about it, there are very few—and this is an anecdote passed to me. There are very few folks that I have ever known or that any of my colleagues have ever known who were liberal arts majors in college who then transferred and changed majors to engineering or chemistry or, you know, biological or physical research kind of activities. But a lot you know that went the other way because it was just such a hard issue. So the approach that has been impressed upon me is if we can gear our education efforts to demonstrate how this excitement in this field can be then brought forward, it is your opportunity then to recruit those who are interested in those fields and sticking with it through college and then through scholarship programs like we are proposing in this legislation, with an objective of looking at those kind of folks as an opportunity to come join the NASA mission objective. That is part of the overall plan, as well.

ATTRACTING PROGRAM MANAGERS TO NASA

Mr. LAMPSON. I think articulating that objective and making sure that they can see a destination, even though we said destination is not necessarily what we want to do right now, is exactly what they want. I think they have to see something that is there that will attract them and keep them involved, that they can see that they will be able to get into a system and stay there, gain the knowledge, the experience of living a project and learning the failures and the difficulties in what the people who developed that project learned because of the one that they worked on before. Which is more or less what George Abbey and many other people have tried to explain to me in every meeting that I have ever sat in at the Johnson Space Center in the 6 years, 5½ years that I have been on this committee.

That is what we have lost. I have wondered if we went back and talked with those people and find out what it was that took them away from staying at NASA. Because it is my opinion, and I have a question here. In Mr. Walker's testimony, he talked about contract management as an Agency challenge, and it is probably the reason why NASA is in so much hot water, both here and in the White House. How is NASA going to find the person with the technical knowledge and the managerial experience to assure the Agency can monitor the contractors' progress against scheduled cost and performance benchmarks to assure the Nation receives the results that it needs?

Johnson is one of the centers, and they are going to need people that have the expertise, not just from the areas of science and engineering, which is what you are saying is going to—we definitely need to work on for the country. Pull them into the interest of going there. But if we do not keep them there for the long haul, give them the opportunity to work on their dream all the way through to its completion, even though there are going to be difficulties in completing some of these things. We are not going to know our way. We are going to get lost. We are going to have cost overruns. We are going to have this, that or the other. How do you find the people and keep them to do that? Is this really going to address that?

Mr. O'KEEFE. Thank you again for the question. I think that is exactly what we are after, and this does not damage, I do not believe, anybody's interest in wanting to be part of this Agency and being part of it for a long period of time. That is inherent in the way all Federal processes developed to this point.

What it augments is an opportunity to assure that given that limited content, smaller number of people we have to choose from today who are coming right out of college, out of graduate schools, doctorals, candidates, etcetera, that you can bring in folks with some experience base from other areas, as we see this retirement go up. Because the bathtub that you have described is there.

We have got a lot of folks who are eligible to retire. We have got a lot of folks that potentially are coming on the front end, and we have got a real mid-level kind of experience based department.

Mr. LAMPSON. We did it to save money because the fact that if we quit hiring along the way, we would spend fewer dollars. What happened was the workforce got older, did get more money and we are now out of money and we are out of knowledge. Not out of it, but we have certainly lost some of each that helps us.

Mr. O'KEEFE. Well, where we are right now, and we put forward a proposal in order to address it. And I think that is the approach we are desperately trying to use, sir.

Mr. LAMPSON. I think also as a part of it, try somehow to work in destination.

Mr. O'KEEFE. Yes, sir. I appreciate your point.

Mr. LAMPSON. Thank you.

Mr. O'KEEFE. Thank you.

WHISTLEBLOWER RIGHTS

Chairman ROHRABACHER. All right. Thank you very much. The Chairman did want to take a second round of questions. I will mention a few things in closing of the hearing and maybe a couple short questions. Mr. Roth, I think that it is not—and I am just assuming this. I try to assume good will on the part of people who disagree, and I cannot imagine that anyone on this panel or in the Executive Branch of Government today is trying to discourage whistleblowers. I mean, there may be people covering their own mistakes in the workforce and come down hard on somebody who is trying to call them on their mistakes. That happens. But, again, I think that type of charge—in reality, I think you will have to admit that issues of security and pay are really paramount to trying to protect the whistleblowers among us.

Mr. ROTH. But I would have to point out that in the Transportation Security Act, it did not provide for whistleblower rights. It gave flexibility to the Deputy Secretary, and he did not agree to give the special counsel an enforceable whistleblower rights.

Chairman ROHRABACHER. This Chairman will support anything that protects whistleblowers in the House or anywhere else throughout the government.

Mr. ROTH. Okay. Because the demonstration project authority——

Chairman ROHRABACHER. Okay. Well, let us——

Mr. ROTH [continuing]. Could give the Administrator the right——

Chairman ROHRABACHER. Let us make sure that we—anything that is proposed to us includes those types of protections. Let me note that I believe that the person who found the shuttle cracks recently was being given a medal, rather than being covered up. I mean, NASA has sought to honor the person who found some problems here. So I do not think that is the mentality that you are up against here at all with them.

Mr. ROTH. Again, I am looking government-wide at where there have been abuses.

Chairman ROHRABACHER. All right. Well, we are talking about——

Mr. ROTH And I have to——

BONUS PAY

Chairman ROHRABACHER [continuing]. NASA today. Let me also note something, Mr. Roth, that you mentioned that some people are going to lose when you have a bonus system. And bonuses are higher than under current law that is being proposed and some people are going to lose. Well, you know, as time goes on, some people lose. There are some people who do not have the skills that are necessary to make the contribution to justify their pay. Because times have changed, and we should not be so stratified and inflexible, pardon the word, to not be able to, you know, make the determination. I am sorry that in order for you to get the same amount of money for what you are doing, you are going to have to make a better contribution than you did 10 years ago when those skills were needed. So some people do lose when you have a dynamic system, and we do need a dynamic system in NASA.

Mr. ROTH. I think you misunderstood me a little bit. As the Comptroller General said, right now, it is difficult because the performance evaluation systems are viewed as not being credible by managers or employees. So if you give a bonus, you know, you are giving it as part of a system that is not viewed as a credible system by either management or the employees. There is a way to get there to credible systems, but we——

NEED FOR INTERNATIONAL AND PRIVATE SECTOR PARTNERSHIP

Chairman ROHRABACHER. I do not think you have any disagreement with the other members of the panel on that. They are trying to reform and restructure the system in a way that makes sense to them. I would recommend they make sure that they keep the employees in the loop and make sure there is a discussion with employees and not just among managers. That is the way to make the system work.

In terms of NASA itself, let me suggest that we are in a situation where there needs to be a restructure. And, again, to have a dynamic situation, you would need to have change in an operation. And there is—the times are changing. I mean, when we went to the Moon, we went on our own, and it was pretty much of a government operation. Today, we need to have more private sector partnership and private sector goals and partnership. We need to make sure there is more cooperation internationally than we have in the past in order to achieve our goals on the Space Station and other goals in the future.

And I think, you know, where we used to have foreign competition, now we need foreign cooperation. So we need a workforce that can work within that context of being able to be flexible and be able to achieve the cooperation that we need to be successful. And let me just note that I am a little disappointed, and let me close with this, and that is for example, I have been on the Administrator now about two potential sources of cooperation with the Government of Quatar and some other Gulf states who might want to contribute to the Space Station efforts and could save us billions of dollars over the long run. I have not been satisfied with the response that I have received by trying to work out that as a potential option, because I am on the International Relations Committee as well, and I have taken a first-hand interest in this.

As well perhaps the response that Mr. Bigalow of Nevada has received from NASA in proposing that he give NASA a storage area for the Space Station and that his company would provide funds for that. So I would hope that NASA shows its flexibility and its willingness to look at new options, as well.

Mr. O'KEEFE. Yes, sir. No. And on the first issue, let me assure you, we will continue to work that. I thought we were there when last we met. But if that is not satisfactory, I will go back and make sure we do——

Chairman ROHRABACHER. I will tell you, it is not satisfactory.

Mr. O'KEEFE. We will work it. You bet. We have got you. And on the second matter, I know we missed each other a week ago. I apologize for that. But let me look into that, as well, in terms of what——

Chairman ROHRABACHER. All right.

Mr. O'KEEFE [continuing]. Answer he did receive and I'll try to respond to that as well.

Chairman ROHRABACHER. All right. So we will be working on these things.

Mr. O'KEEFE. Yes, sir.

Chairman ROHRABACHER. I think the Chairman needs to be a little bit more disciplined in my own Chairmanship, Ms. Jackson Lee, as well as, you know, I am not the most—how do you say—efficient rule manager here, and sometimes that causes problems. So I am going to try and be a little bit more efficient in the job I am doing as the Chair to make sure there is not any conflict and everybody feels they are being treated fairly. With that said, I want to thank all of you. I have got a few closing remarks that I have to say, but before I anything I would like to thank the witnesses again. I certainly would.

Ms. JACKSON LEE. Your graciousness in acknowledging that is appreciated. All of us have deep abiding passions that can sometimes fill over in terms of time, and therefore, accepted in terms of your discipline and certainly recognize the value of my own discipline. As you well know, there are people that are impacted by these issues and therefore, I think in closing, Mr. Gordon has gone—and in closing on his behalf, to a certain extent, you have given us the opportunity.

I will not seek another round, as I understand you might have been happy to do, but I will not seek another round. But I do want to just end on the note that I was making the statement. I do have people in my community that are very concerned, and I hope the flexibility suggests progress and opportunity. And I thank you, Mr. Chairman.

Chairman ROHRABACHER. And they are very concerned about their jobs and their livelihood and their families and you are concerned about them, so that is——

Ms. JACKSON LEE. And being part of that great mission that NASA has. Thank you.

Chairman ROHRABACHER. All right. So with that, I would again thank the witnesses. This hearing actually went on a little bit—actually went on as the length it would have, had every Member been here asking questions for their time. That is sort of the way I justify it. And Subcommittee Members, however, may request additional information for the record, and I would ask other Members who are going to submit written questions to do so within one week of the date of this hearing. With that said, this concludes our hearing and we are adjourned.

[Whereupon, at 12:40 p.m., the Subcommittee was adjourned.]

Appendix 1:

ANSWERS TO POST-HEARING QUESTIONS

ANSWERS TO POST-HEARING QUESTIONS

Responses by David M. Walker, Comptroller General of the United States, General Accounting Office

Coordination Between DOE Science and NSF Research

Q1. You and the NASA Administrator recognized that it would require five to seven years for NASA to transform its workforce and meet the goals of its Strategic Human Capital Implementation Plan. Does NASA's Strategic Human Capital Implementation Plan include an appropriate set of goals and a sound strategy for addressing NASA's human capital challenges? Do you have any recommendations that would strengthen this plan?

A1. Because our work has not entailed a detailed analysis of NASA's Strategic Human Capital Implementation Plan, we are not in a position to report on the appropriateness and soundness of the goals and strategies at this time.[1] However, in response to your question, we observe that NASA's Strategic Human Capital Implementation Plan contains a number of goals and strategies, as well as specific improvement initiatives that the agency is implementing or is committed to implementing. The goals, strategies, and improvement initiatives are organized under five human capital pillars. Furthermore, the plan contains milestones for achieving the improvement initiatives and metrics for assessing their progress.

Until recently, NASA's Strategic Human Capital Plan and companion Strategic Human Capital Implementation Plan were in draft form and were being reviewed by the Office of Management and Budget (OMB) and the Office of Personnel Management (OPM). OMB and OPM approved those plans subsequent to the Subcommittee's hearing—in August 2002. According to NASA, it developed its plans based on OMB's scorecard of human capital standards, OPM's scorecard of supporting human capital dimensions, and our human capital model.[2] The pillars and goals are described in Table 1.

One of the things that I pointed out at the Subcommittee's July 2002 hearing that agencies must do to successfully address their human capital challenges is to ensure that their strategic human capital plans are results-oriented and data-driven. This includes developing appropriate information on the number and location of employees and their key competencies and skills, as well as data on the profile of the workforce, and performance goals and measures for human capital approaches. Further, this data must be used effectively to develop strategies that continually ensure that agencies have the right mix of employees to meet their future needs. A key to success in this area will also be NASA's ability to implement its new financial management system, because it will encompass the integrated workforce planning and analysis system that NASA includes as an improvement initiative in its Strategic Human Capital Plan and plans to implement in 2005: The workforce planning and analysis system is expected to track the distribution of NASA's workforce across programs, capture critical competencies and skills, determine management and leadership depth, and facilitate gap analyses. Currently, NASA has limited capability for personnel tracking and planning, particularly on an agency-wide or program-wide basis. Based on the numerous initiatives NASA will need to undertake, transforming the agency will likely require a multi-year implementation period.

[1] The focus of our work to date has been following up on NASA's actions to respond to its shuttle program workforce challenges.

[2] U.S. General Accounting Office, *A Model of Strategic Human Capital Management,* GAO–02–373SP (Washington, D.C.: Mar. 15, 2002).

Table 1: NASA's Strategic Human Capital Plan Pillars and Goals

Pillar	Goal
Strategic Alignment NASA aligns human capital to support the vision and accomplish the agency's mission and goals.	(1) The agency is well structured organizationally and matches its workforce and workload to support its mission in a safe, effective, and efficient way. (2) Employees understand how what they do and how they do them in support of overall agency goals. (3) Human resources activities clearly, effectively, and efficiently support and enable the agency's mission.
Strategic Competencies NASA recruits, acquires, and retains a diverse workforce with world-class capabilities in strategic competencies needed for all components of its mission.	(4) The agency knows the competencies it needs and the workforce is concentrated in those competencies. (5) NASA recruits, acquires, and retains world-class employees in the essential competencies.
Learning NASA promotes a knowledge-sharing culture and a climate of openness, continuous learning, and improvement.	(6) NASA has a climate of open sharing of relevant knowledge to facilitate best practices, promote personal and professional growth, and avoid failures. (7) NASA actively collects, shares, and utilizes best practices from its successes and lessons learned from its failures.
Performance Culture NASA creates a culture that focuses on results, motivates employees to perform, and ensures fairness in the workplace.	(8) NASA fosters an inclusive climate where employees are valued and treated fairly and with respect, and they feel empowered to make meaningful, relevant, and fulfilling contributions. (9) Equal opportunity and diversity are utilized and valued for their contribution to the agency's mission. (10) The NASA rewards and recognition system acknowledges high-level performance and encourages the behaviors the agency desires in individuals and groups.
Leadership NASA ensures it has leaders who are adaptable; who inspire, motivate, and guide others toward goals; who mentor and challenge the workforce; and who demonstrate high standards of honesty, integrity, trust, openness, and respect.	(11) NASA recruits, selects, hires, and retains a diverse, high performing cadre of leaders who are nurtured through training and development opportunities. (12) Leaders are held accountable for overall performance, at the individual and organizational levels, in maintaining high standards of honesty, integrity, safety and equal opportunity principles.

Source: NASA.

Q2. How effectively is NASA using existing authorities to meet its human capital goals?

Q3. Do you have any recommendations for how NASA can more fully utilize the existing authorities?

Q4. If fully utilized, are the existing authorities sufficient to meet NASA's goals? If not, what goals cannot be met without new or expanded authorities?

Q5. Which proposed new or expanded authorities, if any, are required if NASA is to meet its human capital goals?

A2–5. Our work associated with NASA's human capital challenge has focused primarily on following up on the agency's actions to respond to its shuttle program workforce challenges. At this time, without having performed a more detailed analysis of NASA's human capital plans, we are not in a position to assess NASA's use of existing authorities, the sufficiency of those authorities, and their relationship to its agency-wide human capital goals.

However, we will soon issue a report on human capital flexibilities that could be of significant use to the Subcommittee as it considers NASA's request for additional authorities and flexibilities to recruit, train, and retain critical skills. The report emphasizes that federal agencies should first identify and use the flexibilities available under existing laws and regulations. If additional flexibilities are desired, agencies should develop sound business cases to justify the need for the additional authorities. In this regard, the report discusses agency and union officials' views on the most effective flexibilities for managing their workforces and additional flexibilities that would be most helpful in managing their workforces. It also identifies key practices that agencies should implement to use human capital flexibilities effectively.

According to interviews we held with agency officials at numerous federal organizations and union representatives of several national organizations, work-life programs, such as alternative and flexible work schedules, transit subsidies, child care assistance, and employee assistance programs, are among the most effective human capital flexibilities available in federal agencies for managing the workforce to achieve agency missions and accomplish agency goals. These flexibilities—all of which are currently available to NASA—are effective because they serve as important recruitment and retention tools as employees weigh the balance between their work life and leisure time. In addition, agency and union officials frequently cited monetary recruitment and retention incentives, including recruitment and relocation bonuses for hard-to-fill positions; special hiring authorities, such as limited term appointments and student employment and outstanding scholar programs; and incentive awards to employees for superior job performance and specific accomplishments, including cash and time-off awards.

Categories of additional flexibilities suggested by the officials that would be the most helpful in managing the workforce, if authorized, include (1) more flexible classification and pay approaches, (2) greater flexibility to streamline and improve the federal hiring process, (3) increased flexibility in addressing employees' poor job performance, (4) additional workforce restructuring options, and (5) expanded flexibility in acquiring and retaining temporary employees. These suggestions would provide a starting point for executive branch decision-makers and Congress to consider as they seek to reform federal human capital policies and practices. Key aspects of these additional authorities are included in various legislative initiatives under consideration by Congress. In other cases, however, additional analyses may be needed to ensure that any new authorities are granted and implemented consistent with a focus on program results, merit, and other important federal employment goals.

Our forthcoming report also emphasizes the importance of effectively implementing human capital flexibilities, noting that by more effectively using flexibilities, agencies would be in a better position to manage their workforces, ensure accountability, and transform their cultures to address current and emerging demands. To ensure the most effective use of human capital flexibilities, it is important that agencies

- plan strategically and make targeted investments;
- ensure stakeholder input in developing policies and procedures;
- educate managers and employees on the availability and use of flexibilities;
- streamline and improve administrative processes;
- build transparency and accountability into the system; and
- change the organizational culture.

With regard to NASA, we plan to periodically monitor the agency's human capital planning and implementation efforts and report to Congress as appropriate. We will be pleased to meet with the Subcommittee in this regard.

Q6. What meaningful objectives and measurable goals would you recommend to track progress over the next five years as NASA implements its human capital plans?

A6. It might be better to address this question in terms of the supporting metrics in the human capital plans that NASA has established for assessing its human capital goals. NASA's human capital plans refer to a hierarchy of metrics to assess the agency's progress toward achieving its goals. The plans state that the highest level of metrics are those "critical few" that NASA's senior management and stakeholders care most about and have identified as most appropriate for incorporation into the agency's Government Performance and Results Act-related annual performance and reporting process. Two specific metrics, which the plans state cover a number of goals, have been identified as critical to helping NASA assess its progress toward achieving its human capital aims and assisting it in analyzing root causes of problems and issues. Recognizing that we have not performed a detailed analysis of NASA's human capital plans, these metrics seem meaningful and reasonable to track over the next five to seven years. However, this does not suggest that NASA's commitment to achieving the remaining human capital goals and measures in its human capital plans should be lessened or that the remaining goals and measures should not be tracked. The two critical metrics are:

(1) *NASA's progress in closing gaps in critical competencies.* Using this metric, NASA will be able to determine, on an agency-wide basis, whether the gap between the strategic competencies it has and those it needs is no greater than 10 percent by the end of fiscal year 2004 and that, within the same

period, no gaps exist in those competencies identified as "mission critical." Thereafter, NASA will reduce the strategic competencies gap to no more than five percent within two years and maintain at that level, except there will be essentially no gaps in competencies identified as mission critical.

(2) *Alignment of NASA's human capital strategy with its mission, goals, and organizational objectives.* This metric will allow NASA to determine whether it is achieving an overall combined score that is in the top 10 percent of all scores government-wide on the OPM-administered government-wide survey. This metric will assess NASA's performance relative to whether (a) people have the resources they need to perform their jobs, (b) the performance management system is effective in identifying poor performance and taking steps to improve performance, (c) awards and recognition programs incentivize and reward the behaviors the agency wants to foster, (d) the workforce has adequate opportunities for learning and improvement, and (e) NASA fosters an environment of inclusiveness.

According to NASA's Strategic Human Capital Implementation Plan, the results of the two critical metrics will serve as guideposts for evaluating the overall health of NASA's human capital management effort. NASA believes that they cover directly or indirectly agency progress in addressing all of the areas of concern expressed by internal and external parties regarding the agency's workforce, including:

- identifying and managing competencies needed to carry out the mission and eliminating gaps in mission critical competencies;
- succession planning;
- strategically using existing personnel tools, flexibilities, and technologies;
- differentiating between high and low performers through appropriate incentives and awards and merit-based human resources management; and
- appropriately considering skill mix, technology, electronic government, and competitive sourcing.

Q7. Could NASA meet its goals solely through the use of the proposed alternative personnel system demonstration authority? Are there risks associated with solely using the demonstration authority? Would this present any special challenges for congressional oversight?

A7. NASA's legislative proposal for streamlined demonstration and alternative personnel system authority would provide a mechanism for converting a successful demonstration project,[3] upon approval by OPM, to a permanent alternative personnel system, without the need for special legislation and without any limitation on the number of employees that would be covered by the project as is under current authority. It is our understanding that the NASA Administrator believes that it would be difficult for the agency to meet its goals solely through the use of this proposal. He testified at the July 2002 hearing that the legislative provisions when taken individually would only help NASA deal with its human capital strategic threats to a limited degree. The Administrator stated that when taken together as an integrated package, the provisions would form a strong nucleus in support of NASA's Strategic Human Capital Plan and the President's Management Agenda and would enable NASA to avert a serious human capital crisis.

Without having fully analyzed this proposal in relation to NASA's unique human capital needs, it is difficult at this time to state whether NASA could meet its goals solely through the use of the proposed alternative personnel system demonstration authority. Also, to identify the risks associated with solely using the demonstration authority would require an assessment of the pros and cons of using such an option, particularly if it means bypassing Congress in creating permanent alternative personnel systems and allowing OPM approval without the need for special legislation as proposed by NASA. However, a number of federal agencies have taken advantage of options provided by Congress to test or implement innovative human capital policies, programs, and practices that operate outside the framework of Title 5, United States Code, and regulations under this title, as evidenced by a recent study by The

[3] A demonstration project is a project conducted by OPM, or under its supervision, to determine whether a specified change in personnel management policies or procedures would result in improved federal personnel management (5 U.S.C. 4701(a)(4)). Demonstration projects are conducted directly by OPM or jointly by OPM and the agency or organization. Under Title 5, OPM is permitted to waive certain provisions and regulations under this title to enable agencies or organizations to conduct demonstration projects by experimenting with new and innovative systems (5 U.S.C. 4703(a)).

Partnership for Public Service.[4] The study notes that in most cases these innovations have proven successful in improving the agencies' ability to attract, motivate, or retain a highly qualified workforce while still adhering to the basic principles of a merit-based system. Many of the successful practices were developed through the use of OPM demonstration projects. Pay banding is an approach that has been successfully used in a number of federal organizations, including GAO. It provides the flexibility to offer greater competitive starting salaries and rewards to high-performing employees with financial incentives. As I testified before the National Commission on the Public Service in July 2002, Congress may wish to explore the benefits of providing OPM additional flexibility that would enable it to grant government-wide authority for all agencies to use broadbanding for certain critical occupations and/or allowing agencies to apply to OPM for broadbanding authority for their specific critical occupations. Agencies should be required to demonstrate to OPM's satisfaction that they have modern, effective, and validated performance management systems before they are allowed to use broadbanding. However, the prerogative of congressional oversight must be maintained.

Q8. Could adopting NASA-specific provisions as opposed to taking a government-wide approach lead to a situation wherein scientific and engineering personnel leave other agencies to take advantage of incentives and flexibilities granted to NASA? If so, what would need to be done to prevent such a situation from developing?

A8. When individual agencies are provided flexibilities to address their own strategic human capital needs, it can create a change in the competitive balance for selected personnel. The significance of any related impact would depend on the nature of the flexibilities granted to the agency. For example, expedited hiring authority or early-out and buy-out authority would not have the same impact as higher compensation level authority. Importantly, most federal workers make their employment decisions based on the nature of the job, the mission of the agency, the difference they can make, and the environment within an agency.

Any potential competitive imbalance could be mitigated by limiting the number, amounts and/or term of certain individual flexibilities. Alternatively, government-wide flexibility could be provided in certain areas (e.g., limited-term appointment authority for early-out and buy-out authority and additional compensation authority for selected highly skilled and competitive occupations and positions). This would serve to help level the playing field for attracting and retaining top talent among federal agencies.

Q9. NASA workforce restructuring is being pursued as part of a "results-oriented" management strategy. However, what constitutes a desired result *typically is a function of the policies being pursued by whatever Administration is in power. The civil service system was established to help insulate the federal workforce from politics. How do we ensure that the proposed NASA legislative provisions don't weaken civil service protections and lead to increased politicization of the workforce?*

A9. The Federal Government's existing human capital policies and procedures are based largely on a workforce and labor market that existed in the 1950s. Much has changed in the past 50 plus years and the Federal Government's human capital strategies must be modernized if it is to attract, retain, and motivate top talent in the 21st century. This should include basing federal employment decisions more on the skills, knowledge, and performance of individuals rather than the passage of time and the rate of inflation. However, certain policies should be retained in order to prevent the politicization of the career civil service and to promote certain national goals and policies (e.g., veteran's preference, inclusiveness). Importantly, any human capital system should have appropriate transparency and adequate safeguards (e.g., appeal rights for employees) in order to help ensure that it is efficient, effective, credible, and equitable.

We conducted our work in accordance with generally accepted government auditing standards.

We are sending copies of this report to other interested congressional committees. In addition, the report will be available at no charge on the GAO Website at: *http://www.gao.gov.* If you have any questions about this report or need additional information, please contact Allen Li, Director, Acquisition and Sourcing Management, at (202) 512–4841. Key contributors to this report were Cristina Chaplain, Jerry Herley, Shirley Johnson, and Edward Stephenson.

---

[4] The Partnership for Public Service, *Homeland Security: Winning the War For Talent To Win The War On Terror* (July 31, 2002).

Answers to Post-Hearing Questions

Responses by Sean O'Keefe, Administrator, National Aeronautics and Space Administration

Questions submitted by Chairman Dana Rohrabacher

Q1. How effectively is NASA using existing authorities to meet its human capital goals?

A1. NASA has been aggressive and innovative in using all existing authorities and flexibilities to meet our human capital challenges. NASA was one of the first federal agencies to hire individuals under the Federal Career Intern Program—a program established by Executive Order in 2000 that allows for a more streamlined recruitment process in filling trainee positions in a variety of occupations. NASA also was one of the first agencies to use the new Student Loan Repayment Program to offer an attractive incentive to prospective hires.

NASA's Field Centers are knowledgeable about all available human capital flexibilities and incentives and use them whenever possible and appropriate. For example, the existing recruitment, relocation, and retention bonus authority as well as the superior qualifications appointment authority are incentives long-used by all Centers to attract and retain high quality employees. The use of these incentives has increased in the recent past, but their limitations became apparent as NASA encountered greater difficulty in competing with other employers for needed talent. This led to NASA's legislative proposal to enhance the current recruitment, relocation, and retention bonus authorities to make them more effective financial incentives.

To streamline the lengthy and cumbersome recruitment process that often causes the Agency to lose candidates to the private sector, NASA implemented its Automated Staffing and Recruitment System in FY 2001. This system has simplified and expedited the hiring process by using a computer-assisted rating and referral system and allowing candidates to apply on-line. Despite the significant improvements NASA has seen by using this automated system, existing hiring and selection rules impose unnecessary constraints. The need to further improve these processes led to the proposals for a category rating system and a direct hire authority for positions for which NASA has a critical need or a severe shortage.

The Intergovernmental Personnel Act (IPA) authority is another tool available to federal agencies to meet their human capital needs. NASA has long-recognized the value of the IPA authority as a means of strengthening the workforce by leveraging outside talent from academia and the nonprofit sector; its use within the Agency has been extensive. The legislative proposal to allow IPA assignments to be extended up to six years (rather than four) reflects the importance we place on IPA assignments as a means to sustain the intellectual capital of the Agency. Indeed, this program has proven to be so valuable that we also have proposed creating a NASA Industry Exchange Program, modeled on the IPA, to enable the Agency to engage in mutually beneficial, collaborative ventures with industry to infuse new ideas and perspectives into the Agency, develop new skills within the workforce, and strengthen mission capabilities.

The term appointment authority is another tool that NASA uses to provide greater organizational agility. Term appointments have been very beneficial, particularly in supporting programs and projects of limited duration. Their utility, however, would be enhanced greatly if the appointments could be extended to six years (rather than four) to accommodate the length of some NASA programs and projects. In addition, term appointments would be more attractive to potential candidates if the authority allowed for conversion to permanent positions without the need to go through the same, burdensome hiring process they faced in being appointed to the original term position. This enhancement would encourage a wider pool of high quality candidates to apply for such opportunities. The legislative proposals would provide for these enhancements to the current term appointment authority.

NASA makes extensive use of student employment programs as sources of candidates critical to building a workforce of the future. After a decade of downsizing, we began re-establishing recruitment networks and rebuilding our Cooperative Education Program. We use many other student-oriented programs to develop a future pipeline of talent from which we can draw—such as the NASA Undergraduate Student Research Program and the NASA Graduate Student Research Programs. The recognition of the importance of building the science and engineering pipeline led to our legislative proposal for a "Scholarship for Service" program, which would allow us to support the development of a high-quality science and engineering workforce by awarding scholarships in return for a commitment to work for NASA after graduation.

NASA's most recent buyout authority, which authorized buyouts without losing the corresponding FTE, enabled the Agency to reduce its workforce in areas in which the need for certain skills had diminished and used the FTE's to recruit in areas in which there was a need for different skills. This is an important flexibility for reshaping the workforce to meet changing requirements, and NASA will continue to need it.

The above examples represent just a few of the tools NASA employs to address the Agency's human capital challenges. NASA uses all appropriate human capital authorities and flexibilities, including those that represent long-range as well as short-term strategies. As an indication of the proactive and extensive efforts and achievements we have made in this area, other agencies routinely contact NASA for guidance and assistance as they begin using authorities or programs that already have been tried in this Agency.

Q2. Does NASA have plans to more fully utilize the existing authorities? If so, please explain how these authorities will be more fully utilized?

A2. NASA uses existing human capital authorities and flexibilities to the maximum extent permitted within the legal and regulatory parameters. The Agency uses a wide range of hiring authorities, streamlining the processes associated with them wherever there is discretion to do so. To get the greatest benefit from these authorities, NASA aggressively networks with appropriate institutions and associations to ensure that potential applicants know the Agency's employment opportunities. The emphasis senior management places on effective recruitment resulted in establishing the National Recruitment Initiative in FY 2001 to focus on developing recruitment enhancements, more effective strategies, and improved marketing techniques.

To ensure that effective use of them is not constrained by internal processes or rules, NASA initiated a major "Freedom to Manage" effort this year to review internal (as well as external) barriers and impediments to human capital management. This resulted in delegations of authorities to lower levels, reducing levels of review on some actions, streamlining processes, and making more effective use of automated tools to improve operations.

These efforts will enable NASA to reduce the time it takes to accomplish processes associated with using the existing authorities, and will enable the Agency to use them as effectively as possible, but it does not change the nature or scope of these authorities. Our legislative proposals seek to do that.

Q3. If fully utilized, are the existing authorities sufficient to meet NASA's goals? If not, what goals cannot be met without new or expanded authorities?

A3. The existing personnel authorities and flexibilities are not adequate to address NASA's human capital challenges: lack of depth in critical competencies, skills imbalances, intense competition from the private sector for technical talent, a shrinking science and engineering pipeline from which to draw the talent the Agency needs, and a lack of diversity among some of the Agency's occupations.

Many of NASA's legislative proposals represent enhancements to authorities that already exist; these enhancements are proposed because we have found that the current authorities are not sufficiently broad or flexible to deal with today's challenges. The solutions of the past are not adequate for reshaping and reconfiguring NASA's workforce for the challenging programs and technologies of the future, especially in emerging technical areas like nanotechnology and biorobotics. The limitations of current flexibilities will become even more apparent in the coming years as NASA faces more workforce challenges posed by the continuation of several trends: a reduction in the number of science and engineering graduates, an increase in competition for these shrinking resources from traditional private sector technical employers and the newer competitors for technical talent from other employment sectors (e.g., the banking and entertainment industry), and the potential loss of substantial corporate knowledge to projected retirements of key personnel.

Without additional tools, NASA will continue to face significant strategic threats to our efforts to attract, hire, and retain the workforce we need to fulfill the Agency's mission. These tools are needed to ensure that the Agency has the scientific and technical expertise necessary to preserve the Nation's role as a leader in aeronautical and space science and technology, and a highly competent professional workforce that can address NASA's financial, acquisition, and business challenges.

Q4. Which proposed new or expanded authorities are required if NASA is to meet its human capital goals?

A4. NASA needs all of the authorities submitted in its legislative package. A single human capital tool, flexibility, or authority will not be sufficient to address the diverse strategic threats we face, particularly if we hope to respond quickly and effectively to the dynamics of the marketplace. In certain core areas we need to enhance our ability to recruit and retain highly technical employees. Since this need may encompass all levels of employees—fresh-outs, mid-level, and senior-level—we need different tools and incentives appropriate for these different categories of individuals. Sections 204, 205, 206, 207, and 208 of our proposal are targeted for that need. We also need ways to leverage external expertise more effectively to strengthen NASA's mission and technical capabilities. Sections 202 and 203 are designed to meet that need. At the same time, NASA needs tools to rebalance skills and reshape the workforce. In areas in which we have a diminished need for certain skills we need tools to encourage voluntary attrition so that we can recruit in other areas in which competencies are needed. Sections 209 and 210 will enable NASA to accomplish this. NASA must continue to encourage students to pursue science and engineering careers to ensure a steady pipeline of technical talent for the future. Our proposed Scholarship for Service Program, submitted in the NASA Science and Technology Career Enhancement Act of 2001, will support this goal. Finally, we need to be able to try out new flexibilities, more quickly and efficiently than we now can, in response to future program and organizational changes as well as labor market shifts. Section 211 would help us with this challenge.

Our provisions represent an integrated set of human resources flexibilities that will allow the Agency to address the multiple challenges it faces without in any way affecting the important protections or entitlements in areas of merit principles, equal employment opportunity, employee benefits, veterans preference, and labor relations.

Q5. What meaningful objectives and measurable goals would you recommend to track progress over the next five years as NASA implements its human capital plans?

A5. In developing the Agency's Strategic Human Capital Plan (SHCP) and accompanying Strategic Human Capital Implementation Plan (SHCIP), NASA was influenced by a number of factors: internal and external reviews touching on the Agency's human capital issues; the Administration's emphasis on human capital through the President's Management Agenda; legislative branch interest in Government wide human capital challenges; and current nationwide and internal trends that threaten NASA's ability to maintain a highly skilled workforce. Using OMB, OPM, and GAO standards and tools as guidance, NASA established an overarching human capital architecture structured around five "Pillars"—Strategic Alignment, Strategic Competencies, Learning, Performance Culture, and Leadership, the last of which provides the foundation for all the others. Each pillar was further defined in terms of high-level goals and strategies. Once the architecture was in place, a team of senior NASA managers did an assessment—or "gap analysis"—of the Agency's current strengths and weaknesses in human capital management.

While the Agency already has in place a wide variety of programs, initiatives, and activities to address recruitment, retention, and training and development of the workforce, NASA's internal assessment of the Agency's current state of human capital management highlighted nine major focus areas—or key "improvement initiatives"—that merit special emphasis. These are organized under the five pillars. Together, their successful implementation will greatly enhance the Agency's ability to effectively manage its human capital and maintain its preeminence as a world-class organization with a highly motivated, skilled, productive and innovative workforce—and enhance realization of a "One NASA."

Action plans have been developed for each of the nine initiatives and we are in the process of establishing success indicators to help determine whether the initiatives, once implemented, are achieving their intended objectives. These success indicators will form part of a hierarchy of measures to help the Agency assess the overall effectiveness of human capital management. At the highest level of measures—those senior Agency managers and stakeholders care most about—are those to determine 1) NASA's progress in closing gaps in strategic competencies and 2) alignment of NASA' human capital strategy with its mission, goals, and objectives. (The attachment summarizes this architecture.)

Q6. Could NASA meet its goals solely through the use of the proposed alternative personnel system demonstration authority? Are there risks associated with solely using the demonstration authority? If so, how would NASA propose to minimize these risks?

A6. NASA cannot meet all of its human capital goals solely through the use of the proposed personnel demonstration project authority. The Office of Personnel Management (OPM) has advised that buyout authorities or modifications to the early retirement authority cannot be implemented through a demonstration project. These two authorities are essential to NASA's efforts to rebalance skills and reshape the workforce. They provide a means to encourage targeted attrition in areas in which the need for certain skills has diminished so that the Agency can recruit and reshape a workforce that is better aligned to current and future mission needs.

The remaining legislative proposals could, theoretically, be implemented through a demonstration project authority. Implementing a demonstration project involves coordination and negotiation with OPM, employees, and labor organizations. Although we are confident that these proposals will continue to be endorsed by OPM and will be supported by the NASA workforce, there is no *guarantee* that we would be successful in our efforts to incorporate all of them, in their current form, in a future demonstration project.

A more realistic and significant concern, however, relates to the time and resources involved in implementing and evaluating a demonstration project. Our proposal (Section 211) greatly streamlines the process, but it remains, nevertheless, very time-consuming and labor-intensive. NASA has major human capital challenges now, which require immediate attention to prevent their further escalation. It is unlikely that a demonstration project could be implemented in less than a year from the date legislation is passed authorizing the "streamlined" authority. Our legislative proposals could be implemented expeditiously, enabling NASA to address today's challenges before they become tomorrow's problems.

Implementing a demonstration project and evaluating its results represent a significant workload, and the workload increases commensurate with the number and complexity of provisions in the project. A number of the proposals in our legislative package reflect flexibilities that have been tested in demonstration projects of other federal agencies, or provided to agencies through specific legislation (e.g., The Internal Revenue Service reform legislation). Since these are "proven" practices, rather than untested departures from current rules, pursuing them through another demonstration project that would take more than a year to implement unnecessarily delays our efforts to address our significant, and immediate, human capital challenges.

Q7. How is NASA working with the university community to expand the pipeline of aerospace scientists and engineers?

A7. NASA believes that it is extremely important to engage the university community in efforts to expand the supply of qualified scientists and engineers. NASA's education program has recently established two primary strategies to achieve our goal motivating students to pursue careers in science, technology, engineering, and mathematics (STEM). Both these strategies address the Agency's human capital needs and one—*increase the number of individuals entering the STEM workforce*—specifically targets the university community. To implement these strategies NASA plans to rely on partnerships with appropriate education organizations, particularly universities.

NASA implements a coordinated and comprehensive set of programs that provide support to a diverse population of students preparing at universities for STEM careers. These programs include undergraduate, graduate, and post-graduate university support fellowships and scholarships.

The largest and most visible of NASA's education programs is the Space Grant College and Fellowship program. Space Grant provides both graduate and undergraduate fellowships through consortia of universities in all 50 states, the District of Columbia, and Puerto Rico. The Space Grant Directors represent a mechanism for working with universities to expand the pipeline. For example, through our recently announced workforce development awards, NASA selected 45 consortia in the Space Grant university network to receive funding for aerospace workforce development. A total of $3.56 million is being awarded in response to proposals submitted by university consortia. The consortia were selected based on a competitive evaluation of their plans to enlarge and enhance the "pipeline," or resource pool, of higher education graduates and faculty who stay connected to or become involved with NASA as employees, contractors or principal investigators.

Q8. Why are NASA's stipends for graduate students working on aerospace studies almost 20 percent less than NSF stipends for comparable work?

A8. NASA, like other agencies, determines stipend levels for programs such as the Graduate Student Research Program (GSRP) based on several factors, including a comparison of stipends at other mission agencies, perceived "competitiveness" of the program, and available budget.

In FY 2002, NASA raised the award for the GSRP to $24,000 after an analysis of other federal fellowship programs indicated that the agency was no longer competitive. The award includes a stipend of $18,000, and $6,000 in student/university allowances, which can be used to help defray tuition expenses. Additional stipend increases are planned, based on continuing analyses and as program budgets permit.

Q9. Is this low stipend a disincentive for graduate students to become involved with NASA research?

A9. We do not think so. Our graduate programs, like all of our student programs, continue to be "over subscribed"—that is, we have more applicants than we can accept, and they are of a high caliber.

It is difficult to compare NASA GSRP stipends to graduate stipends at NSF, as a mission agency like NASA has an intrinsic value that NSF cannot match—our mission, facilities, and people. More than half of our GSRP students actually spend time at a NASA center, working with a research mentor. The remainder is working with NASA Principal Investigators at their respective universities.

Questions submitted by Ranking Minority Member Bart Gordon

Q1. Have you used your authority under Section 203(c)(2) of the Space Act to establish the entrance grade for scientific and engineering personnel at a level two grades higher than the General Schedule? If not, why not? If so, how has this affected recruitment success?

A1. Yes. NASA has used this authority to establish an entrance grade of GS–7 for scientific and engineering candidates who would otherwise be eligible for appointment at the GS–5 level. This authority is used in connection with a NASA-specific qualification standard requiring the candidates have specific academic credentials or work experience that would be appropriate for NASA aerospace technology work.

This flexibility is used along with the various federal-wide incentives and flexibilities—including tools such as accelerated promotions, special salary schedules for engineers and other occupations, and recruitment bonuses. These recruitment incentives have been effective in past years, but they are no longer adequate in today's competitive job market. New incentives and flexibilities are needed to compete successfully with the private sector to attract a world-class workforce. We believe that an effective way to address our recruitment challenge is to pursue flexibilities that reflect enhancements of well-established and successful compensation policies. For example, the legislative provision for enhanced recruitment, relocation, and retention bonuses would represent a recruitment incentive that would be effective, while remaining consistent with federal-wide qualification and classification standards.

Q1a. Do you believe that the Space Act authority cited above would allow you also to increase the step-level that could be offered?

A1a. Appointing an individual at a step above the minimum rate currently is permitted under the superior qualifications appointment authority.

Q2. With respect to the proposed Scholarship for Service program and the statutorily authorized Federal Student Loan Repayment program:

- *Does NASA participate in the Student Loan Repayment program?*
- *If not, why not?*
- *If so, how does the program work at NASA, and what are its effects on recruitment?*

A2. Yes, NASA participates in the Student Loan Repayment program, using it as one of several recruitment tools to attract college graduates to NASA. Along with recruitment bonuses and offers of salaries above the minimum rate, student loan repayments comprise an attractive incentive to recent college graduates who may face substantial debts upon graduation. Since this incentive is one of a variety of tools available to recruiters, it is not offered in every case, and is subject to funding limitations. Each agency is required by regulation to develop an implementation plan before offering student loan repayments. At NASA this program, as most others, is highly decentralized, with each Center using a basic template to devise a program to meet its individual needs and resources. Since this program is quite new, and implementation had to await the development of the required plans, a relatively small number of loan repayments have been offered to date. According to our latest data, about a dozen loan repayments have been approved. We anticipate that loan repayments will become an integral part of the incentive package offered to prospective employees.

Q2a. With respect to the proposed Scholarship for Service program and the statutorily authorized Federal Student Loan Repayment program:

- *Do you have any quantitative data as to whether a "Scholarship for Service" program would provide more or less incentive than a Student Loan Repayment program in attracting graduates to NASA?*

A2a. Since one program is in its infancy and the other still on the drawing board, we have no quantitative data to predict their relative impact on attracting graduates to NASA. The two programs actually are quite different in intent and approach. The results naturally will be different, though they will complement each other. The "Scholarship for Service" program strives to encourage students currently in college to study science, engineering, and technology curricula. With the service requirement, NASA naturally hopes that a number of the participants will come to work for NASA and remain beyond their service requirements.

In contrast, the Student Loan Repayment program targets those who have already graduated or for whom graduation is imminent, and is not limited to a particular area of study. Its specific purpose is to entice graduates to NASA by offering a financial incentive. As mentioned above, this incentive can be meshed with other tools to craft a recruitment package that will attract a candidate of choice to come to work for NASA.

Q2b. With respect to the proposed Scholarship for Service program and the statutorily authorized Federal Student Loan Repayment program:

- *Would Scholarship for Service employees be allowed to join collective bargaining units?*
- *Would a Scholarship for Service participant be hired as a Term employee or a Permanent employee?*

A2b. During the required period of service, scholarship recipients could be hired under a special temporary employment authority designed for that purpose, or they could be hired as regular permanent employees. This employment decision would rest with NASA management, based on staffing needs, and would reflect the desires and intentions of the individual. Service under a temporary appointment would satisfy the service requirement, and some graduates will be ready to move on to new experiences after completing their employment obligation. Others will find a permanent place in NASA's workforce; naturally, this result benefits both the individual and the Agency. The type of appointment, and whether the bargaining unit's charter covers the employee's position would determine an employee's inclusion in a collective bargaining unit. Some charters cover employees on time-limited appointments, while others limit membership to permanent employees. The employee's status as a "scholarship recipient" has no direct impact on his or her eligibility to join a collective bargaining unit.

Q3. Have you conducted any formal surveys of the NASA workforce on issues related to the proposed human capital provisions? If so, when, and on which provisions?

A3. Over the past several years a number of broad reviews have been conducted within the Agency that have addressed workforce issues relevant to NASA's proposed human capital legislation.

In March 2000, the Office of Management and Budget (OMB) and NASA engaged in a joint study to develop a strategy for meeting NASA's human capital challenges for the next decade. The team conducted an extensive benchmarking process that included interviews and focus groups with individuals at most NASA Centers including Center Directors, Senior Managers, Line Managers, new hires, and union representatives. This comprehensive review was completed in September of 2000, resulting in workforce issues and challenges and recommended actions to address the identified challenges. The recommendations resulting from this study in part led to the development of the specific human capital provisions in NASA's proposed legislation. Among the legislative proposals recommended by the joint review team were: targeted buyouts to address skills imbalances without the loss of FTEs; extensions of IPA assignments to six years; establishment of an Industry-Exchange program; expansion of the critical pay authority; conversion of term employees to permanent; and streamlined hiring authorities.

In 2001, NASA established a National Recruitment Initiative (NRI) to develop Agency-wide recruitment strategies to attract and hire a highly technical S&E workforce. As part of its data collection effort, the NRI study team conducted interviews at seven NASA Centers with directors of science and engineering, human resource directors and chiefs of employment, recruiters, equal opportunity staff, university affairs officers, hiring managers, and new/recent hires. The interviews provided valuable information in shaping NASA's current legislative proposals by calling attention to several strategic threats now facing the Agency: the inability to offer salaries competitive with private industry; anticipated loss of critical skills over the next five years due to retirements; difficulty in attracting a diverse candidate pool; and the potential loss of quality candidates due to the current slow and cumbersome hiring process.

The Aerospace Safety Advisory Panel's Annual Reports for 2001 and 2000 have noted that workforce concerns continue to be a focus of the Panel. These reports contained several recommendations that influenced the provisions of NASA's human capital legislative proposals, including recommendations to provide more effective incentives to retain employees with critical skills in specific technical areas and to use appropriate incentives when necessary to recruit experienced and fresh-out employees.

In addition to these NASA-focused surveys, we monitor national trends that could affect NASA's ability to maintain a workforce with the talent it needs to accomplish its mission. These include trends associated with math and science performance among students, the science and engineering "pipeline," general labor market conditions, projected demands for specific skills, and surveys on perceptions of employment in the federal sector. The National Science Board's Science and Engineering Indicators 2002 Report, for example, contains important analyses and projections that are relevant to this Agency's human capital management strategies.

Q4/5. Are there statutory precedents for NASA's proposed legislative provisions (Sec. 211):

- *to eliminate the limit on the number of personnel that could be included in an alternative personnel system demonstration?*

A4/5. There are statutory precedents for eliminating the limit on the number of personnel that could be covered in a demonstration project.

(1) *Department of Defense Science and Technology Reinvention Laboratories Demonstration Projects*—The current authority to conduct personnel demonstrations is found in Title 5, United States Code, Chapter 47. The National Defense Authorization Act for FY 1995 (Public Law 103–337) extended the authority to conduct personnel demonstrations to the Secretary of Defense, with the approval of the Director of the Office of Personnel Management. Under this statute, the Secretary is authorized to conduct demonstrations in DOD laboratories designated as Science and Technology (S&T) Reinvention Laboratory sites. The statute further stipulates that the Title 5, Chapter 47 limitation on the number of personnel that could be included in a demonstration would not apply to any of these DOD laboratories. Currently there are eight distinct demonstration projects being conducted under this authority across the Air Force, Navy, and Army. The DOD demonstration authority was further revised by Public Law 106–398, which eliminated the requirement for OPM approval of any S&T Reinvention Laboratory demonstration project.

(2) *Internal Revenue Service Demonstration Project*—The IRS Reform Legislation (Public Law 105–206) authorizes the Secretary of the Treasury to implement a demonstration project for the IRS subject to the general provisions of Title 5, United States Code, Chapter 47, but without limitation on the number of personnel that could be included in the demonstration.

Q4/5a. Are there statutory precedents for NASA's proposed legislative provisions (Sec. 211):

- *to make an alternative personnel system permanent without a demonstration and without Congressional approval?*

A4/5a. Generally, "alternative personnel system" refers to a personnel system that provides for the waiver of Title 5 (or a rule or regulation prescribed under that title) or provides an authority not in current law, rule or regulation. If an agency seeks innovations to its personnel system that require such waivers or new authorities, it must either pursue the changes through legislation or obtain approval to test the innovations under an existing demonstration project authority. If the agency implements an alternative system through enabling legislation, the language itself will specify whether a demonstration phase is needed, specify the extent to which Congressional review/approval of the system is needed, and indicate if the system is permanent. If the agency implements innovations under the current demonstration project authority, it cannot make them permanent unless special legislation is enacted to do so.

The Federal Aviation Administration's alternative personnel system was implemented as a permanent system without a demonstration phase. Section 347 of the Department of Transportation Appropriations Act of 1996 (Public Law 104–50) directed the FAA to develop and implement a new personnel management system by January 1, 1996 to address the unique demands of the Agency's workforce and provide for greater flexibility in the hiring, training, and compensation of personnel. The law exempted the new personnel system from the provisions of Title 5, United States Code, with specific exceptions (e.g., provisions relating to whistleblower protection, veterans preference, anti-discrimination). The FAA established an alternative personnel system as directed. An alternative personnel system was made permanent without undergoing a demonstration phase and the *specific* provisions of the system were not subject to Congressional review and approval prior to implementation. Nevertheless, the establishment of the alternative personnel system was done at the direction of Congress.

There is another method by which agencies can waive Title 5 provisions that does not require testing the innovations in a demonstration phase and does not require Congressional approval. This approach involves granting agencies the authority to replace specific provisions of Title 5 with the more flexible rules of Title 38 of the United States Code. Title 38 represents a separate personnel system for certain health care occupations in the Veterans Health Administration hospitals. This system, established in 1946 to facilitate the rapid staffing of veterans hospitals at the end of World War II, contains flexibilities not found in Title 5. In 1990, The Federal Employees Pay Comparability Act (Public Law 101–509) gave the Office of Personnel Management authority to extend specific provisions of Title 38 personnel and pay flexibilities to other agencies to use in hiring and compensating individuals for General Schedule health care positions. This is now codified in Section 5371 of Title 5. Agencies using these flexibilities are not required to undergo a demonstration or "test" phase first. However, the authority to establish these alternatives to Title 5 laws, rules, and regulations was provided by Congress through Public Law 101–509. It should be noted, however, that this method of introducing personnel flexibilities is narrowly defined in the sense that it simply replaces specific Title 5 hiring and compensation rules with the rules from Title 38, and the flexibilities are used only in connection with health care positions.

Q6. In addition to workforce, the Strategic Resources Review was to address infrastructure requirements at the NASA Centers.

- *Does NASA have an excess of centers, facilities, or other physical infrastructure, or is the current level of infrastructure appropriate for the Agency's requirements over the next decade?*
- *If there is excess, what specific infrastructure have you identified for elimination, and what is your schedule for eliminating it?*

A6. The now completed SRR identified some assets that will be closed, mothballed, or excessed (e.g., a thermal vacuum chamber, two Gulfstream 1 aircraft) as noted in the final report delivered to Congress in August 2002. NASA is continually evaluating its facility needs and other physical infrastructure as part of the President's Management Agenda and NASA's new Vision & Mission. As part of this ongoing facility and physical infrastructure evaluation, recommended changes are reviewed during NASA's internal budget formulation process. Any change to the Agency's facilities or other physical infrastructure will be provided to Congress for their consideration in future budget submissions.

Q7. Do you believe that it would be useful for NASA and the Congress to have a review that assesses the future roles and missions of each NASA Center, what facilities should be closed and what facilities should be added or upgraded, what NASA activities could be contracted out, and what future workforce levels should be?

- *If so, when do you plan to initiate such a review, and by what date would you expect to complete it?*
- If not, why not?

A7. Given the recent completion of the SRR, we do not see a need at this time for a special review of the roles and missions of each NASA center. The President's Management Agenda, in particular its Competitive Sourcing initiative, provides the direction for regularly determining whether particular activities are more efficiently conducted inside of NASA or by the private sector. We have no particular goals for future workforce levels or outsourcing—such levels should be determined through competition in performing NASA's missions.

Q8. Could adopting NASA-specific provisions as opposed to taking a government-wide approach lead to a situation wherein scientific and engineering personnel leave other agencies to take advantage of incentives and flexibilities granted to NASA? If so, what would need to be done to prevent such a situation from developing?

A8. NASA's legislative provisions represent tools and flexibilities designed to: encourage students to pursue careers in science and engineering; enable the Agency to compete more successfully with the private sector in attracting a world class and diverse workforce; reshape the existing NASA workforce to address the current skills imbalances; and leverage expertise from academia and the private sector. Few of the incentives or flexibilities represented in these legislative proposals would provide any immediate benefit to current federal employees who might transfer to NASA from other agencies.

The provisions regarding the *Intergovernmental Personnel Act authority* and the *Industry Exchange Program* involve exchange assignments between current NASA employees and non-federal organizations. Neither provision would provide an inducement for an employee in another federal agency to seek employment in NASA since there would be no reason for the employee to assume that he or she would ever have an opportunity to participate in such an exchange. (Generally, NASA has fewer than 15 NASA employees participating in the IPA program at any given time, and participation among NASA employees in an Industry Exchange Program, if enacted, may be just as limited.)

The provisions regarding *voluntary early retirement* and the *buyout authority* are tools to reshape the existing NASA workforce; they are not recruiting incentives that would attract current federal employees from other agencies to NASA.

The *hiring authority* for severe shortage/critical needs positions and the *category rating authority* are tools that would streamline and improve the external hiring process; they would not apply to filling NASA jobs through transfers of other federal employees. NASA's current policies and processes for hiring transfer eligibles from other government agencies remain unchanged.

The provisions regarding *term appointments* would not encourage permanent employees from other government agencies to seek employment with NASA since permanent employees generally would not wish to convert to term status. A very similar provision has been in place for several years in most of the DOD Science and Technology Reinvention Laboratory Demonstration Projects and the IRS.

The provisions to increase the maximum *rate of pay for NASA Excepted* (NEX) employees is unlikely to have any impact on other agencies since the NEX hiring authority generally is used in appointing individuals from the private sector.

The *critical pay authority* would have no impact on other agencies since its use is limited to retaining NASA employees or appointing individuals who are not federal employees. It cannot be used in connection with transfers of employees from other agencies.

The *recruitment and relocation bonus* authorities are financial incentives that NASA would use to recruit a quality workforce. Since such bonuses could be offered to federal employees of other agencies as well as "new hires," they could be used to incentivize an employee to leave his or her current agency to accept an offer with NASA. However, we anticipate that our use of the bonus authority will be focused primarily on attracting external candidates to NASA. Due to funding realities and the need to use bonuses judiciously, we anticipate that guidelines for their use will be patterned after existing rules limiting them to filling "difficult to fill" positions. For NASA, these positions are typically filled with fresh-out graduates or individuals with specialized expertise found in the private sector.

The authority to conduct demonstration projects has been available to agencies for over two decades and has not been circumscribed during this time due to any perception that employees are leaving their current agencies to seek employment in an agency under a demonstration authority. NASA's legislative proposal for a *streamlined demonstration project* does not change any of the underlying principles of the demonstration authority or revise the prohibitions against waiving certain statutes or regulations (such as those pertaining to employee benefits, merit system principles and prohibited personnel practices, ethics rules, and veterans preference). So, NASA would not be implementing any flexibilities that would not be available to any other agency under a demonstration authority. NASA's proposal simply streamlines the *processes* involved in implementing a demonstration project, and removes the 5,000-person limit on the number of individuals that can be included in a project.

Questions submitted by Representative Sheila Jackson Lee

Q1. Considering the difficulty NASA is having in recruiting the best and brightest young people, and the dearth of minorities in senior level NASA positions—would you support the idea of a program which would bring promising graduate students from historically disadvantaged communities into NASA for short periods of time? In this time they could shadow NASA employees and witness the facilities at work during two launches, and thus, hopefully, get excited about building a career furthering the mission of NASA.

A1. Attracting the Nation's students, especially minority students, into the fields of science, mathematics and engineering is vital to our country's continued success in these fields and an important part of NASA's mission "to inspire the next generation of explorers." One of the trends we are facing is the reduction in the number of graduates in science, math, and engineering. The resultant loss of talent for our workforce of the future poses one of our greatest challenges. Increasing minority representation in our workforce is also hampered by this trend. One of our tools for meeting this challenge is the Harriett G. Jenkins Predoctoral Fellowship Program (JPFP). This is a national, competitive three-year program for minority graduate students that provides each fellow a NASA mentor and coordinates visits to their mentor's site. Annually, all fellows compete and 10 per cohort are selected for a six-week research internship program conducted at a NASA Center. Currently, forty multi-talented fellows are in this Program. Five of the fellows have inquired about making application to become an astronaut. At least one fellow is being considered for "Graduate Student of the Year" through the U.S. Black Engineer of the Year Program. Through 2005, the program should contribute 11 Masters and 12 Doctorate graduates to the NASA and science, math and engineering workforce.

In addition every one of our Field Centers offers student internship programs for minority students. For example, the NASA Scholars at Morehouse, Spelman, Florida A&M and University of New Mexico Universities participate in intern programs at all NASA Centers; the Society of Hispanic Professional Engineers internship program and Strategic Preparedness in Advancing Careers in Engineering, both at Stennis Space Center; Summer Institute in Engineering and Computer Applications Program and Public Service Intern Program at Goddard Space Flight Center; the NASA Scholars Program at the Johnson Space Center; and the Navigator Minority Outreach Program at the Jet Propulsion Laboratory. Minority students are also eligible for the large number of internship and fellowship programs offered by NASA to all students, including the Space Grant Scholarship Program.

We are convinced that the opportunity to come on site and participate first hand in the activities taking place at our Field Centers will indeed help inspire that next generation of explorers. For this reason the Johnson Space Center is looking into the feasibility of developing a short term internship program with Texas Southern University whereby some of its graduate students may visit JSC, work with a NASA scientist or engineer, and, if timing permits, witness the excitement at JSC during a shuttle mission.

Q2. Does NASA have a role to play in promoting Homeland Security? Are there technologies and expertise at NASA, which could be used to enhance cyber and satellite security? If yes, what would be the best method of making such technology available to the Department of Homeland Security, and facilitating cooperation, consultation, or contracts?

A2. While NASA does not directly combat terrorism, security activities at NASA's field centers and headquarters defend the agency and its mission against terrorism, thus enabling it to fulfill its role as the premier space exploration and scientific research agency, unimpeded by threats or adversarial influences. NASA has therefore made critical security enhancements to its ten field centers and headquarters, the White Sands Test Facility, and Goldstone Tracking Station. These enhancements reduce the vulnerability of NASA Centers to security threats and protect its personnel, critical assets, and mission essential infrastructure. In this way, NASA directly contributes to the security of cyber and satellite assets.

Our indirect role in and support of Homeland Security lies in NASA's numerous partnerships with other federal agencies that have a direct role in the protection of the United States, and can take advantage of NASA's technology, research, and unique capabilities. NASA's research applications offer a variety of unique approaches to managing the many threats to homeland security. NASA has innovative programs in aviation safety, agriculture monitoring, human heath and community disaster preparedness, coastal monitoring, and biological detection and analysis being implemented at virtually every Center across the Agency. NASA has established partnerships with NIH, DOD, NIMA, FAA, TSA and several other agencies—we are using these relationships to leverage NASA's contribution to Homeland Security. NASA's most immediately available work with homeland security applications falls into three broad categories, Sensors, Air Travel and Remote Sensing:

Sensors: Whether managing the closed environments of spacecraft and the International Space Station, searching for water on Mars or making detailed observations of our own planet; NASA's long experience with sensors has led to the development of a number of small, low power, rugged monitoring and detection systems that are being shared with other agencies. Of particular interest has been NASA's miniature mass spectrometer, which can detect a variety of chemicals accurately, i.e., chemical weapons. A variety of other technologies, like the electronic nose, deep-ultraviolet imaging and DNA based sensors are also available for collaborative development. When combined in suites or deployed on semi-autonomous robots, these technologies have tremendous potential.

Air Travel: We have ongoing collaborative tasks through the TSA and the FAA to make aircraft safer and harder to crash, and to improve the Nation's air traffic control systems. We are utilizing our information technology resources to assist in passenger threat assessment, and provide air traffic management decision support; for example, identification of rogue aircraft, minimization of risk to other aircraft and ground targets. Our aeronautics researchers are focusing on 'hardening' aircraft systems: reducing the explosive risk in all systems; improving communication between aircraft and ground; and deploying intelligent systems which can detect, alert and counter dangerous flight deviations. We have developed a special partnership with the FAA for development of new sensor capabilities in detection of explosives in baggage as well as biological and chemical weapons and their components.

Remote Sensing: Working with NIMA and other DOD agencies, NASA's remote sensing and observation resources are being used now to safeguard the Nation and fight the war on terrorism abroad. We help provide critical mapping, weather and terrain data wherever it is needed, domestically or in a foreign battlefield. Additionally, we have briefed NSA and OSTP on the global observations and prediction capabilities from NASA enabled science and technologies. Meta-data analysis, or, the synthesis of many kinds of information like maps, weather, real time measurements from field sites for use in a command center environment, is under active development at NASA

While our ability to fund and staff homeland security tasks is limited by statute, we have taken steps to incorporate information relevant to homeland security into our technology database and our interagency relations office continues to serve as a networked point of contact both inside and outside NASA. This will ensure that the widest possible number of key people will be aware of NASA technology with homeland security applications.

Q3. If NASA does have a role in Homeland Security, and if NASA does have important information systems and satellites that may be targets for hackers or terrorists, does NASA's lack of monitoring and control of contract workers make us vulnerable to inside attack?

A3. No, NASA has mechanisms in place for monitoring and control of contract workers at NASA.

NASA continues to enhance its control and access procedures for all its personnel, so as to minimize vulnerabilities to its systems at all Centers. Since August of last year, NASA has embarked on a new Security program and initiative to eliminate or reduce all vulnerabilities while minimizing the effect it would have on the Agency's mission.

Are we as an agency vulnerable to inside attack? No organization or agency is 100 percent safe from such an attack. The solution is to minimize the threat and damage through rapid and effective alert/response/mitigation; and appropriate screening, security, and education.

ANSWERS TO POST-HEARING QUESTIONS

Responses by Mark D. Roth, General Counsel, American Federation of Government Employees, AFL–CIO

Questions submitted by Chairman Dana Rohrabacher

Q1. What workforce issues does NASA face and to what extent are these issues unique to NASA?

A1. NASA is somewhat of a microcosm, reflecting issues and trends that many executive branch agencies face with regard to recruiting, motivating, and retaining qualified federal employees. The most important single issue has to do with contracting out. NASA has for more than two decades pursued a strategy of contracting out not only work performed competently and efficiently by its own workforce, but also virtually all new work and expansions of work. The predictable result is that the skill base and morale of the existing workforce has eroded, and young and mid-career professionals interested in aerospace engineering and astrophysics look to government contractors, not NASA, for employment. The problems are quite pronounced at NASA, but the Department of Defense and other civilian agencies face the same consequences of excessive contracting out.

Q2. Are some of these issues unique to NASA because of the type of workforce and skills employed at NASA? Is NASA's uniqueness a relevant factor in planning its workforce strategy?

A2. AFGE does not believe that NASA's difficulties with recruitment and retention are unique. Thus NASA does not need a separate or unique "workforce strategy." NASA and other federal agencies need to pay competitive salaries, offer opportunities for career development and growth in a merit system, and stop contracting out existing and new work.

Q3. How much of NASA's workforce problem is the result of broader trends in education and the aerospace industrial base?

A3. The private contractors who have cannibalized NASA's work and workforce do not face the same "human capital" crisis that NASA faces. This is because they offer their employees competitive pay and meaningful career development, as well as the opportunity to do their work without the constant distraction of privatization. In addition, they get all new government work, and automatically get expansions on existing projects. Contracting out and failure to pay competitive salaries are the reason for NASA's workforce problems.

Q4. What specific skill areas represent the most critical and urgent needs with NASA's workforce?

A4. AFGE cannot answer this question with any certainty.

Q5. What changes in size and skill mix should NASA consider over the next five years as an increasing amount of the space and aeronautics work is contracted out to industry?

A5. AFGE opposes treating increasing contracting out as inevitability. NASA currently lacks adequate in-house ability to provide taxpayers with oversight of its contracts, which are increasing rapidly in terms of both cost and quantity. NASA is unable to tell either Congress, the Office of Management and Budget (OMB), or the American people how much it spends on contracts, how many contract workers it supports with taxpayer funds, or what scope of government work is being performed by those contractors. NASA has failed to train or retain an adequate in-house capacity to avoid placing taxpayers at the mercy of monopoly, for-profit contractors. At a minimum, NASA needs an adequate in-house scientific and engineering capacity performing work of sophistication equivalent to its contractors so that adequate oversight can occur. NASA, like every other executive branch agency, needs to rebuild and expand its acquisition workforce for contract management and administration as well.

Questions submitted by Ranking Minority Member Bart Gordon

Q1. Could adopting NASA-specific provisions as opposed to taking a government-wide approach lead to a situation wherein scientific and engineering personnel leave other agencies to take advantage of incentives and flexibilities granted to

NASA? If so, what would need to be done to prevent such a situation from developing?

A1. AFGE strongly opposes the adoption of NASA-specific solutions to what is a government-wide set of problems. NASA's difficulties with recruitment and retention, and its inability to monitor effectively its massive and costly contracts, are the direct result of misguided *government-wide* policies. The failure to fund or implement the Federal Employees Pay Comparability Act (FEPCA) which mandates the payment of competitive salaries to federal employees is a government-wide phenomenon. The mandate to privatize and contract out up to half of all jobs arbitrarily designated as "commercial" is a government-wide mandate enforced by OMB. These two policies have conspired to create NASA's workforce problems and reversing these two policies will be necessary first steps in solving those problems. If one executive branch agency begins paying higher salaries than other agencies for a specific occupation or position, there will be nothing preventing exodus from the agency with low salaries. NASA is not the only federal agency employing scientists and engineers; the Department of Defense and other civilian agencies face similar problems.

Q2. NASA workforce restructuring is being pursued as part of a "results-oriented" management strategy. However, what constitutes a desired result *typically is a function of the policies being pursued by whatever Administration is in power. The civil service system was established to help insulate the federal workforce from politics. How do we ensure that the proposed NASA legislative provisions don't weaken civil service protections and lead to increased politicization of the workforce?*

A2. The Bush Administration is at war with its workforce, and seems intent on creating a spoils system that moves money and jobs away from what they perceive as a unionized public sector toward their political patrons in the contracting community. The types of managerial discretion given to political appointees, and the enhanced authorities given to management generally, is not counterbalanced by any expansion in collective bargaining rights in the legislation. On the contrary, federal employees would lose the opportunity they currently have to have their views heard by their elected representatives in Congress since Congress would no longer have final authority over the range of federal salaries. Politicization would also result from an abandonment of the principle of equal pay for equal work, and the setting of pay on an individual by individual basis, rather than by classification or position description.

Q3. NASA is proposing a sweeping set of workforce-related legislative provisions. What level of consultation has NASA or the Administration had with the employee unions on these provisions?

A3. NASA management has not once contacted AFGE to describe or discuss the workforce related changes it seeks. Likewise, no representative of the Bush Administration has contact AFGE to discuss these workforce related legislative changes.

Q4. Your testimony is critical of the Administration's proposed NASA–Industry exchange program (proposed section 203) that would allow private sector employees to work at NASA and NASA employees to work in industry. What are your specific objections to the proposal? Are there any alternative approaches that you would suggest be considered?

A4. NASA's failures to conduct adequate oversight and monitoring of the billions of taxpayer dollars it spends on private contracts are legion. As the Bush Administration continues to pursue a policy of mandatory contracting out of as many as 850,000 federal jobs it arbitrarily designates as "commercial" in nature, allowing favored contractors intimate access to remaining in-house operations puts taxpayer interests in peril. Assuming that OMB's privatization mandates include some measure of public-private competition (even though thus far the President's "competitive sourcing initiative" is competitive in name only; the vast majority of government work that is contracted out is privatized without giving federal workers the opportunity to compete in defense of their jobs), taxpayers' interests are best served if the in-house team isn't forced to share with its competitors all information on its operations. AFGE does not oppose allowing federal employees to have limited details into private firms so that taxpayers gain the benefit of any insight or innovation the private sector is willing to provide.

Q5. NASA's proposed section 204 to authorize direct hiring for positions in "critical needs" and "severe shortage" categories would appear to give NASA discretion to define those categories and to decide how many positions would be exempt from the current merit-hiring process. What are your views on the proposed pro-

vision? If you do not support it, is there an alternative approach you would suggest for dealing with critical hiring needs?

A5. There is every reason to expect that discretion to hire directly and avoid merit system principles will be abused. That is why the civil service system currently limits such discretion. It is very difficult to disprove an assertion of a critical need or severe shortage. Likewise, it is very difficult to prove that an individual hired outside the merit system was hired due to political connections. Taxpayers have the right to expect that those hired by the government have obtained their jobs based on what they know, rather than whom they know. AFGE believes that the slowness in hiring reported anecdotally is more a function of inadequate staffing in agencies' human resources departments that the hiring system itself. Human Resources offices were affected disproportionately in the excessive and unfocused downsizing of the 1990s. Downsizing combined with contracting out have rendered many agencies with administrative systems that hamper the merit system hiring process. On-the-spot impulsive hiring that opens the door to political favoritism is not a solution to this problem.

Appendix 2:

Additional Material for the Record

Lewis Engineers and Scientists Association

IFPTE LOCAL 28 21000 Brookpark Road Cleveland, OH 44135-3191

(216)433-5623 Phone (216)433-5627 FAX

COMMENTS OF INTERNATIONAL FEDERATION OF PROFESSIONAL & TECHNICAL ENGINEERS (IFPTE), AFL-CIO, LOCAL 28, NASA GLENN RESEARCH CENTER, CLEVELAND, OHIO

HOUSE SPACE & AERONAUTICS SUBCOMMITTEE HEARING "NASA WORKFORCE & MANAGEMENT ISSUES"

JULY 18, 2002, 2318 RAYBURN HOUSE OFFICE BUILDING WASHINGTON, DC

IFPTE, Local 28, Lewis Engineers & Scientists Association (LESA), located at the NASA Glenn Research Center in Cleveland, Ohio would like to thank the members of the Space & Aeronautics Subcommittee for allowing us to submit testimony to be included as a part of the record for today's hearing. In particular, IFPTE, Local 28 would like to extend special appreciation to Chairman Rohrabacher and Ranking Member Bart Gordon.

IFPTE, Local 28 represents over 1600 Federal Bargaining Unit Employees holding positions classified as General Service (G.S.) Scientists, Engineers, Professional, Administration, Clerical and Technical.

Our statement for the record is as follows:

General Concern:

President Bush has identified Competitive Sourcing as a key element of the President's Management Agenda. While we support the notion of continually assessing the cost and quality of services performed both within the government as well as those functions obtained commercially, numerous features of the current Competitive Sourcing process are grounds for enormous concern. As it is presently proceeding, this process will not insure that the taxpayer's interest of cost and quality are adequately served, and at the worst will promote long term and possibly irreparable damage to important public institutions such as the National Aeronautics and Space Administration (NASA). Many of these concerns are reflected in the recent final report of the Commercial Activities Panel (CAP), David Walker, Comptroller General of the United States, Chairman.

Specific areas of concern are as follows:

1. The use of numerical goals for competitive sourcing is disturbing. As stated by the CAP, "The success of government programs should be measured by the results achieved in terms of providing value to the taxpayer, not the size of the in-house or contractor workforce." *The arbitrary numerical quotas imposed by the Bush Administration, and the accompanying*

requirement to submit Competitive Sourcing Plans reflective of these numerical quotas is driving untethered and arguably damaging plans to eliminate wholesale functions within government institutions. Such plans constitute knee-jerk reactions at best, and are not underpinned by adequate alignment with long-term Agency missions, and are not supported by any cost/benefit studies whatsoever. Since NASA is presently engaged in a top-to-bottom reassessment of its strategic priorities and supporting workforce requirements, it is clearly inappropriate to have such quotas force the outsourcing of functions while issues such as legitimate benefit and the identification of core government functions remain outstanding.

The rush to initiate competitive sourcing activities in the absence of legitimate cost/benefit analyses is already resulting in the displacement of present in-house government functions, involving questionably ethical or legal practices. This is particularly visible for unique functions that require significant periods (e.g. years) of on-the-job experience and proficiency, and thus cannot be obtained by the direct substitution of contractors, regardless of what claims of skills and expertise are touted by the offeror. As noted in the CAP report, "there should be adequate safeguards to ensure that activities, entities, or functions are not improperly unbundled as a means to come under the ceiling to avoid competition."

2. A clear and complete assessment of the extent to which NASA functions are already outsourced is lacking. Although the individual Field Centers have recently been asked to submit data regarding services obtained from the commercial sector, the guidelines for this exercise were vague and inconsistently followed, such that an accurate accounting of the balance between in-house and outsourced functions is impossible to determine. It should be noted, however, the estimates from the Field Centers concur that the Agency presently out sources in excess of 87% of its total budget. *No assessment has been undertaken by NASA regarding the balance of in-house and outsourced functions necessary to perform the mission for whom the Agency is responsible. In a recent telecon regarding the classification of "commercializable functions" under the FAIR Inventory, NASA Code R Administrator Dr. Jerry Creedon (then the Director of the NASA-Langley Research Center) observed that outsourcing goals undertaken over the past decade have already driven the Agency to "the brink of unmanageability."*

3. The use of FAIR Inventory data on "commercializable functions" as a basis for achieving the Bush Administrations arbitrary numerical outsourcing goals spells disaster. *By its nature, the FAIR Inventory process attempts to inappropriately reduce the nature of government personnel and functions to overly simplistic tabular data. The net result fails to capture numerous facets that are fundamental in properly assessing the nature and extent of inherently government functions.* As examples, the Inventory contains no provisions reflecting the need for unbiased and non-vested mission analyses. This would inevitably lead to a corruption of honest and independent assessments of technical directions and approaches. The Inventory also fails to address the critical need for intellectual and infrastructure continuity. A commercial source could easily elect to abandon its role as service provider in a critical area should it be viewed down the line to serve the financial interests of its stockholders. Further, the inventory fails to reconcile the important coupling between certain competencies, such as the requirement for a critical mass of in-house research and technical staff (the so-called "smart buyers") necessary to insure appropriate program content and direction, and to do so in an independent and unbiased

fashion. For example, the Inventory unilaterally classifies all research functions as "commercialiazable," reflective of no requirement whatsoever to retain a suitable number of in-house technical personnel for the reasons described above.

4. The lack of fair and open competition between the government and commercial sectors remains rampant. As noted in the CAP report, *"the Panel heard concerns that federal employees risk being at an unfair advantage in a competition with the private sector. In particular, concern was raised that federal employees' efforts to put together a truly "most efficient organization" have sometimes been hamstrung by lack of cooperation or unwillingness to commit resources on the part of management. This alleged lack of cooperation included a failure to keep employees informed about the process, a lack of support for innovative suggestions from employees, and unwillingness to make the capital investments necessary to allow the in-house plan to win the cost comparison, and a failure to give employees the tools to conduct a thorough reengineering analysis of their work and to propose a competitive plan."* Indeed, there continues to be a consistent unwillingness throughout NASA to include members of the workforce at any level in the drafting or review of Competitive Sourcing Plans. It is impossible to imagine, on this basis, that a fair and honest assessment of the perceived benefits and any ensuing outsourcing initiatives would, in fact, ever be realized.

5. Outsourcing initiatives to date have been accompanied by a lack of accountability. Lessons learned are essentially absent, and clear and consistent examinations of the benefits gained (either cost or quality) remain generally unavailable. Again, as reflected in the CAP report, "in the federal procurement system today, there is a common recognition that a cost-only focus does not necessarily deliver the best quality or performance for the government... In this sense, the Circular [A-76] is an anomaly in the federal procurement process in that it effectively inhibits consideration of important, non-cost factors." *Before embarking on further and potentially damaging downsizing of our in-house government infrastructure, it is critical to demand a thorough assessment of the benefits [or lack thereof] that have resulted from previous outsourcing initiatives.*

APEX
TP
9780160696510
76901945 — 49